Ali Mobasheri, Carolyn A. Bondy,
Kelle Moley, Alexandrina Ferreira
Mendes, Susana Carvalho Rosa,
Stephen M. Richardson, Judith A.
Hoyland, Richard Barrett-Jolley,
Mehdi Shakibaei

Facilitative Glucose Transporters in Articular Chondrocytes

Expression, Distribution and Functional Regulation of GLUT Isoforms by Hypoxia, Hypoxia Mimetics, Growth Factors and Pro-Inflammatory Cytokines

With 34 Figures

Ali Mobasheri

School of Veterinary Medicine and Science
University of Nottingham
Sutton Bonington Campus
Loughborough, Leicestershire
LE12 5RD, United Kingdom

e-mail: ali.mobasheri@nottingham.ac.uk

Carolyn A. Bondy

National Institute of Child Health
and Development
National Institutes of Health
Bethesda MD20892, USA

Kelle Moley

Department of Obstetrics and Gynecology
Washington University School of Medicine
4911 Barnes-Jewish Hospital Plaza
St. Louis MO 63110, USA

Alexandrina Ferreira Mendes
Susana Carvalho Rosa

Faculty of Pharmacy and Centre for
Neurosciences and Cell Biology
University of Coimbra
3004-517 Coimbra, Portugal

Stephen M. Richardson
Judith A. Hoyland

Tissue Injury and Repair Group,
School of Clinical and Laboratory Sciences,
Faculty of Medical and Human Sciences,
University of Manchester, Manchester,
M13 9PT
United Kingdom

Richard Barrett-Jolley

Department of Veterinary Preclinical Sciences
Faculty of Veterinary Science
University of Liverpool
Liverpool L69 7ZJ
United Kingdom

Mehdi Shakibaei

Musculoskeletal Research Group
Institute of Anatomy
Ludwig Maximilians University Munich
80336 Munich, Germany

ISSN 0301-5556
ISBN 978-3-540-78898-0   e-ISBN 978-3-540-78899-7

Library of Congress Control Number: 2008929622

9 8 7 6 5 4 3 2 1

springer.com

Reviews and critical articles covering the entire field of normal anatomy (cytology, histology, cyto- and histochemistry, electron microscopy, macroscopy, experimental morphology and embryology and comparative anatomy) are published in Advances in Anatomy, Embryology and Cell Biology. Papers dealing with anthropology and clinical morphology that aim to encourage cooperation between anatomy and related disciplines will also be accepted. Papers are normally commissioned. Original papers and communications may be submitted and will be considered for publication provided they meet the requirements of a review article and thus fit into the scope of "Advances". English language is preferred.

It is a fundamental condition that submitted manuscripts have not been and will not simultaneously be submitted or published elsewhere. With the acceptance of a manuscript for publication, the publisher acquires full and exclusive copyright for all languages and countries.

Twenty-five copies of each paper are supplied free of charge.

Manuscripts should be addressed to

Prof. Dr. F. **BECK,** Howard Florey Institute, University of Melbourne, Parkville, 3000 Melbourne, Victoria, Australia
e-mail: fb22@le.ac.uk

Prof. Dr. F. **CLASCÁ,** Department of Anatomy, Histology and Neurobiology,
Universidad Autónoma de Madrid, Ave. Arzobispo Morcillo s/n, 28029 Madrid, Spain
e-mail: francisco.clasca@uam.es

Prof. Dr. M. **FROTSCHER,** Institut für Anatomie und Zellbiologie, Abteilung für Neuroanatomie,
Albert-Ludwigs-Universität Freiburg, Albertstr. 17, 79001 Freiburg, Germany
e-mail: michael.frotscher@anat.uni-freiburg.de

Prof. Dr. D.E. **HAINES,** Ph.D., Department of Anatomy, The University of Mississippi Med. Ctr.,
2500 North State Street, Jackson, MS 39216–4505, USA
e-mail: dhaines@anatomy.umsmed.edu

Prof. Dr. N. **HIROKAWA,** Department of Cell Biology and Anatomy, University of Tokyo,
Hongo 7–3–1, 113-0033 Tokyo, Japan
e-mail: hirokawa@m.u-tokyo.ac.jp

Dr. Z. **KMIEC,** Department of Histology and Immunology, Medical University of Gdansk,
Debinki 1, 80-211 Gdansk, Poland
e-mail: zkmiec@amg.gda.pl

Prof. Dr. H.-W. **KORF,** Zentrum der Morphologie, Universität Frankfurt,
Theodor-Stern Kai 7, 60595 Frankfurt/Main, Germany
e-mail: korf@em.uni-frankfurt.de

Prof. Dr. E. **MARANI,** Department Biomedical Signal and Systems, University Twente,
P.O. Box 217, 7500 AE Enschede, The Netherlands
e-mail: e.marani@utwente.nl

Prof. Dr. R. **PUTZ,** Anatomische Anstalt der Universität München,
Lehrstuhl Anatomie I, Pettenkoferstr. 11, 80336 München, Germany
e-mail: reinhard.putz@med.uni-muenchen.de

Prof. Dr. Dr. h.c. Y. **SANO,** Department of Anatomy, Kyoto Prefectural University of Medicine,
Kawaramachi-Hirokoji, 602 Kyoto, Japan

Prof. Dr. Dr. h.c. T.H. **SCHIEBLER,** Anatomisches Institut der Universität,
Koellikerstraße 6, 97070 Würzburg, Germany

Prof. Dr. J.-P. **TIMMERMANS,** Department of Veterinary Sciences, University of Antwerpen,
Groenenborgerlaan 171, 2020 Antwerpen, Belgium
e-mail: jean-pierre.timmermans@ua.ac.be

200
Advances in Anatomy, Embryology and Cell Biology

Editors

F. Beck, Melbourne · F. Clascá, Madrid
M. Frotscher, Freiburg · D.E. Haines, Jackson
N. Hirokawa, Tokyo · Z. Kmiec, Gdansk
H.-W. Korf, Frankfurt · E. Marani, Enschede
R. Putz, München · Y. Sano, Kyoto
T.H. Schiebler, Würzburg
J.-P. Timmermans, Antwerpen

List of Contents

Abstract

Articular cartilage is a unique and highly specialized avascular connective tissue in which the availability of oxygen and glucose is significantly lower than synovial fluid and plasma. Glucose is an essential source of energy during embryonic growth and fetal development and is vital for mesenchymal cell differentiation, chondrogenesis, and skeletal morphogenesis. Glucose is an important metabolic fuel for differentiated chondrocytes during postnatal development and in adult articular cartilage and is a common structural precursor for the synthesis of extracellular matrix glycosaminoglycans. Glucose metabolism is critical for growth plate chondrocytes which participate in long bone growth. Glucose concentrations in articular cartilage can fluctuate depending on age, physical activity, and endocrine status. Chondrocytes are glycolytic cells and must be able to sense the concentration of oxygen and glucose in the extracellular matrix and respond appropriately by adjusting cellular metabolism. Consequently chondrocytes must have the capacity to survive in an extracellular matrix with limited nutrients and low oxygen tensions. Published data from our laboratories suggest that chondrocytes express multiple isoforms of the GLUT/SLC2A family of glucose/polyol transporters. In other tissues GLUT proteins are expressed in a cell-specific manner, exhibit distinct kinetic properties, and are developmentally regulated. Several GLUTs expressed in chondrocytes are regulated by hypoxia, hypoxia mimetics, metabolic hormones, and proinflammatory cytokines. In this multidisciplinary text we review the molecular and morphological aspects of GLUT expression and function in chondrocytes and their mesenchymal and embryonic stem cell precursors and propose key roles for these proteins in glucose sensing and metabolic regulation in cartilage.

1
Introduction

The provision of nutrients and oxygen to synovial joints is essential for the physiological and load-bearing functions of articular cartilage and the homeostatic control of metabolism within chondrocytes, its resident cells (Mobasheri et al. 2002c; Mobasheri et al. 2006). The transport of nutrients (i.e., glucose, other hexose and pentose sugars, amino acids, nucleotides, nucleosides and water soluble vitamins such as vitamin C) into articular chondrocytes is essential for the synthesis of collagens, proteoglycans, and glycosaminoglycans by chondrocytes (Clark et al. 2002; Goggs et al. 2005; McNulty et al. 2005; Mobasheri et al. 2002a). There are numerous biological mechanisms by which nutritional factors might be expected to exert favorable influences on cartilage function and pathophysiological events in disease processes including osteoarthritis (McAlindon 2006). A decade ago, very little was known about nutrient transport in chondrocytes, particularly the transport of glucose, related sugars, and water-soluble vitamins, which are essential for the synthesis of glycosaminoglycans by chondrocytes. Glucose is a crucial nutrient for cartilage function in vivo as it is for many other tissues and organs. However, it has always been assumed that glucose is important for the in vitro cultivation of chondrocytes, ex vivo maintenance of cartilage explants, and cartilage tissue engineering procedures. No-one had actually studied the molecular mechanisms responsible for the uptake of glucose and glucose-derived vitamins such as vitamin C until the early 1990s when Bird and co-workers and Hernvann and colleagues studied the kinetics of glucose transport by chondrocytes and synovial fibroblasts in the presence and absence of proinflammatory cytokines (Bird et al. 1990; Hernvann et al. 1992, 1996). By the late 1970s it was well established that ascorbic acid supplementation was essential for maintaining sulfated proteoglycan metabolism in chondrocyte cultures and growth plate metabolism, hypertrophy, and extracellular matrix (ECM) mineralization (Gerstenfeld and Landis 1991; Schwartz and Adamy 1977; Schwartz et al. 1981). Studies by Otte and Hernvann related the importance of glucose as a metabolic substrate (Otte 1991) and emphasized the fact that glucose uptake is stimulated by catabolic cytokines in chondrocytes (Hernvann et al. 1992) and that stimulated glucose uptake is inhibited by anti-inflammatory cortisol (Hernvann et al. 1992, 1996). The last 8 years have seen significant progress in this area of connective tissue research although perhaps not compared to advances in our knowledge in other tissues. Therefore, the time has arrived in chondrocyte physiology and cell biology for recapitulation. Novel information gained about the physiological roles of transporters, particularly those involved in glucose uptake for essential metabolic and biosynthetic reactions in cartilage, and their regulation by hypoxia, growth factors, and cytokines, may contribute to a better understanding of the altered molecular and cellular mechanisms in cartilage pathologies such as osteoarthritis (OA) and osteochondritis dissecans (OCD). The aim of this monograph is to review the recently published information on the expression, distribution, and regulation of isoforms of the facilitative glucose transporter family (GLUTs) in articular chondrocytes and their stem cell precursors. We also explore the physiological implications of the functional regulation of these proteins by hypoxia, hypoxia mimetics, growth factors, and proinflammatory cytokines. This text also reviews our recent studies aimed at understanding the process of metabolic regulation in chondrocytes. In particular we discuss the functional expression of K_{ATP} channels in chondrocytes and their putative roles in the regulation of extracellular glucose and intracellular ATP sensing (Mobasheri et al. 2007). Glucose transport and metabolism in chondrocytes play key roles in the biology and physiology of articular cartilage. Glucose uptake is actually a major limiting step in glucose utilization by chondrocytes. Future progress in dealing with degenerative joint disorders such as OA, OCD, and related joint disorders will be highly dependent on a better understanding of the unique nutritional requirements of chondrocytes. Research is currently underway to determine if nutrient transport systems in chondrocytes and synoviocytes and related metabolic mechanisms and signaling pathways may offer suitable targets for modulating the behavior and biosynthetic activity of articular chondrocytes. A clearer knowledge of chondrocyte nutrition and the regulation of transport systems responsible for nutrient uptake in chondrocytes in health and disease may reveal underlying metabolic disturbances that are directly responsible for cartilage degradation in OA and other arthropathies. This knowledge will lead to new approaches and novel therapies to prevent and treat degenerative joint disease and enhance the approach for the discovery and design of drugs capable of modifying degenerative joint diseases.

1.1
Nutrient Sensing: A Fundamental Property of All Living Cells

All living cells must be able to regulate their metabolism when faced with nutrient fluctuations in their extracellular environment. Nutrient sensing is defined as a living cell's ability to recognize and respond to fuel substrates such as glucose. Each type of metabolic fuel used by living cells requires a distinct and carefully regulated uptake and utilization pathway involving transport, regulatory, and accessory molecules. In order to conserve valuable resources a cell will only produce biomolecules that it requires. These requirements may change when cells are involved in different activities such as division, proliferation, differentiation, and apoptosis. The quantity and type of metabolic fuel that is available to a cell will determine the complement of enzymes it needs to express from its genome for efficient utilization of the available nutrient. Some metabolic fuels are also important structural precursors for the synthesis of other biochemicals. Glucose is an example of a metabolic fuel and a structural substrate for the synthesis of glycoproteins and glycoconjugates. Specific receptors on the cell membrane's surface designed to be activated in the presence of specific fuel molecules communicate to the cell nucleus by means of biochemical signaling cascades. This mechanism allows cells to maintain awareness of the available nutrients in their environment in order to adjust their metabolism to utilize the available substrate molecules more efficiently.

Nutrient sensing is critical to the survival of all prokaryotes and eukaryotes: studies in plants (Rolland et al. 2002), yeast (Forsberg and Ljungdahl 2001), and bacteria (Gilmore et al. 2003) have demonstrated that these organisms are able to sense and respond to changes in extracellular carbon and nitrogen metabolites. In eukaryotic and mammalian cells the physiological maintenance of nutrient and metabolite homeostasis is crucial for many fundamental cellular functions (Rolland et al. 2001)—including cell division, proliferation, differentiation, excitability, secretion—and it is also critical to cell fate, senescence (Nemoto et al. 2004), and apoptosis (Martens et al. 2005). Cells generally adapt to alterations in the extracellular concentrations of any given nutrient by regulating its transport rate across the plasma membrane (and subsequently its storage and metabolism). Such adaptation is essential for numerous subcellular functions and may involve transcriptional control of transporter genes and cell surface sensors.

1.2
Glucose: The Universal Energy Currency

Glucose is the principal carbon and energy source for all living cells and is the most abundant monosaccharide on the planet (Saier et al. 1999a). Glucose is an important energy currency on land and in the oceans where plants and phytoplankton actively fix carbon dioxide into carbohydrates. At both unicellular and multicellular levels the quantity of available glucose can fluctuate considerably and living organisms must be able to sense the amount of glucose available to vital tissues and organs and respond appropriately by adjusting their metabolic rate. Altering the expression of glucose transporter genes is one of the major effects that changes in glucose concentration will have on living cells. Lessons learned from studies on glucose sensing and signaling mechanisms in plants and yeast have significantly enhanced our understanding of how cells from higher organisms sense and respond to glucose. In plants and microorganisms glucose and other sugars function as nutrients and as 'signaling molecules,' exerting transcriptional control over many nutrient transporter genes and altering the subcellular distribution of nutrient transporters (particularly the glucose transporter or GLUTs—see subsequent sections) (Alpert et al. 2002; Kumar et al. 2004; Shin et al. 2004). Extracellular sugars also alter mRNA and protein stability. Accordingly, the cells, tissues, and organs of all living organisms have evolved sophisticated molecular mechanisms for sensing extracellular glucose. Invariably, when this control is lost in mammals, glucose homeostasis is compromised and what follows is diabetic pathology which can affect a whole range of tissues and organ systems.

Glucose is an important substrate for all mammalian cells and provides a source of readily available energy for cellular metabolism. Sugars also provide carbon skeletons for the biosynthesis of other macromolecules such as proteins, lipids, nucleic acids, and complex storage polysaccharides (glycogen). Furthermore, hexose sugars are building blocks of glycoproteins such as proteoglycans which in addition to their role as structural components of the ECM fulfil adhesive and informational functions. Even modest changes in glucose concentrations in the microenvironment of cells may impair anabolic activities mediated by growth hormone and growth factors and catabolic activities mediated by proinflammatory cytokines and matrix metalloproteinases. Therefore, the steady supply and transport of physiological levels of glucose is critical for cell viability and ECM synthesis.

In the following section we briefly discuss the well-established example of blood glucose regulation by feedback regulation using the coordinated actions of the hormones glucagon and insulin produced by pancreatic α and β cells in mammals.

1.3
Regulation of Glucose Metabolism by Feedback Regulation

Glucose homeostasis in the mammalian circulatory system is maintained by 'feedback regulation' (Halter et al. 1985; Kahn and Porte 1988; Leibiger et al. 2002; Porte and Kahn 1991). Feedback regulation is a homeostatic control mechanism that uses the consequences of a biological process to regulate the rate at which the process occurs. The feedback loop for glucose homeostasis keeps blood glucose at a steady set point of 5 mM. The actual glucose concentration within blood represents the net flux of glucose entering and leaving the bloodstream (Casey 2003). Glucose is delivered to the blood from the liver following ingestion of carbohydrate, from the breakdown of glycogen (the storage form of glucose), and from endogenous glucose synthesis. Glucose is removed from the blood by a number of organs and tissues, notably the brain, skeletal muscle, liver, and adipose tissue (Casey 2003). Despite the complexity of the regulation of blood glucose, the fundamental homeostatic control mechanism is quite simple and consists of several key elements: (1) glucose-responsive α and β cells in the pancreas; (2) the hormones insulin and glucagon; (3) insulin- and glucagon-responsive tissues that store and utilize glucose (i.e., skeletal muscle, cardiac muscle, and adipose tissue) or release or re-synthesize glucose by gluconeogenesis (i.e., liver); and (4) the glucose set point concentration as the regulated parameter. When blood glucose rises after a meal, insulin is released by pancreatic β cells to stimulate uptake and utilization of glucose by insulin-responsive tissues thus restoring normal plasma glucose. Conversely, when blood glucose falls below 5 mM (i.e., during fasting), glucagon is released to stimulate glycogen breakdown and glucose release from glucose stores. When the pancreatic β cells fail to produce sufficient insulin after a meal or when peripheral insulin-responsive tissues lose their ability to respond to normal concentrations of insulin, the result is diabetes. Pancreatic hexokinase and the GLUT2 facilitative glucose transporter are considered to be important components of the glucose-sensing apparatus in pancreatic β cells (Waeber et al. 1995), and the sodium-dependent glucose transporter SGLT1 has been proposed to be a component of the glucose sensors in renal, intestinal, and neuroendocrine cells (Gribble et al. 2003; Wood and Trayhurn 2003). Although much less is known about glucose sensing in the brain, a number of putative glucose sensors have been identified in hypothalamic neurons. These include the glucagon-like peptide-1 receptor (GLP-1R), glucose transporter isoform 2 (GLUT2), hexokinase, and ATP-sensitive K^+ channels (K_{ATP} channels) (Alvarez et al. 2005; Arluison et al. 2004a, 2004b; Evans et al. 2004). In the following section we briefly discuss the process of glucose sensing before introducing the readers to the structure and function of articular cartilage and reviewing the relevant literature on glucose transporters in chondrocytes and their stem cell precursors.

1.4
Glucose Sensors: Lessons Learned from Other Organisms

Studies in bacteria, yeasts, and plants suggest that glucose acts as a 'signaling nutrient' and a 'hormone' (Moriya and Johnston 2004). Glucose is a major source of carbon and energy for many bacteria, yeasts, and plants. Studies in *Saccharomyces cerevisiae* suggest that glucose transporters in the yeast cell membrane are key components of the glucose-sensing pathway. Since extracellular glucose needs to be determined by yeast cells it would make sense to place the sensors on the cell surface in contact with the extracellular environment. The rate of glucose utilization in yeast is dictated by the abundance of glucose transporters in the plasma membrane. Molecular analysis of sugar transporters in *Saccharomyces cerevisiae* has revealed the existence of a multigene family of sugar carriers (Bisson et al. 1993) and many of the genes encoding sugar carriers are putative glucose sensors due to their plasma membrane localization. Indeed the yeast genome encodes at least 18 hexose (Hxt) transporters (Ozcan and Johnston 1999). Yeast cells can detect extracellular medium glucose and respond to generate subcellular signals that affect gene expression to regulate the cellular response to alterations in the concentration of extracellular glucose. These alterations occur at the mRNA level by activating or repressing genes that encode proteins involved in glucose transport and metabolism (Johnston 1999). Discussion of yeast glucose transporters and the G-protein coupled receptor-mediated sensing of extracellular glucose in yeast is beyond the scope of this review and has been covered elsewhere (Forsberg and Ljungdahl 2001; Ozcan and Johnston 1999).

2
Articular Cartilage: Structure, Function, and Pathophysiology

In the following two subsections we discuss the structure and function of articular cartilage and give a broad background to the pathophysiology of arthritis before discussing the links between nutrition and arthritis and how diet affects skeletal development. We also discuss how nutrition might predispose humans and animals to arthritis or protect us from it. We then discuss the links between nutrition and joint disease in the context of our studies on glucose transporters in chondrocytes.

2.1
Normal Articular Cartilage Structure and Function

Articular cartilage is an avascular, aneural and alymphatic connective tissue designed to distribute mechanical load and to provide a wear-resistant surface to articulating joints (Buckwalter and Mankin 1998b) (Fig. 1A–D). It consists of a tough and mechanically resilient ECM that occupies more than 90% of tissue volume when it is fully hydrated. Chondrocytes are the resident cells of cartilage and are the only cells found interspersed within the ECM (Aigner et al. 2006b; Carney and Muir 1988). Chondrocytes occupy between 1% and 10% of the total tissue volume (depending on the animal species studied, its developmental stage and the age of the animal). The ECM is composed of collagens, predominantly collagen type II, which give tensile strength to the cartilage and restrain the osmotic swelling pressure caused by the hydrophilic aggregating proteoglycans (aggrecan); thus aggrecan hydrates the cartilage, thereby giving it viscoelasticity and allowing it to withstand compressive force (Dudhia 2005; Maroudas 1976; Vachon et al. 1990). This ability to hydrate and resist compressive force is aided by sulfated glycosaminoglycans (GAGs), such as the abundant chondroitin sulfate and to a lesser extent keratan sulfate, along with nonsulfated GAGs such as hyaluronan, all of which attach to their specific domains on the aggrecan molecule to form the aggregates (Watanabe et al. 1998). While aggrecan comprises about 35% dry weight of the total protein content of the ECM, there are also small nonaggregating proteoglycans known as small leucine-rich repeat proteoglycans (SLRPs) (Dudhia 2005). The SLRPs are decorin, biglycan, fibromodulin, and lumican, and their core proteins allow them to interact with various collagens and growth factors in the ECM thereby giving them roles in regulating fibrillar collagen formation (Font et al. 1998; Vogel et al. 1984), strengthening the linkage of collagen to other cartilage macromolecules (Wiberg et al. 2003), limiting access of proteolytic enzymes to the cleavage site on the collagen molecule (Geng et al. 2006; Sztrolovics et al. 1999), and modulating chondrocyte metabolism via regulating growth factor availability (Hildebrand et al. 1994; Roughley 2006). Within the ECM, there are also glycoproteins

Fig. 1 Micrographs of porcine articular cartilage stained with hematoxylin and eosin clearly showing the zonal organization of cartilage and morphology of chondrocytes in the surface (**A**), middle (**B**) and deep (**C**) zones. The confocal rotation series (**D**) illustrates a group of five chondrocytes from the middle zone of bovine metacarpal phalangeal cartilage. Isolated porcine chondrocytes maintain their morphology in suspension culture (**E**) but rapidly dedifferentiate in two-dimensional culture and begin to take on fibroblastic characteristics (**F**). The electron micrograph shown in **G** illustrates chondrocytes in three-dimensional organoid culture isolated from the limb buds of 7-day-old mice. [Reproduced from Mobasheri et al. 2002c, with copyright permission of *Histology and Histopathology* (http://www.hh.um.es/) and Jiménez-Godoy, S.A.]

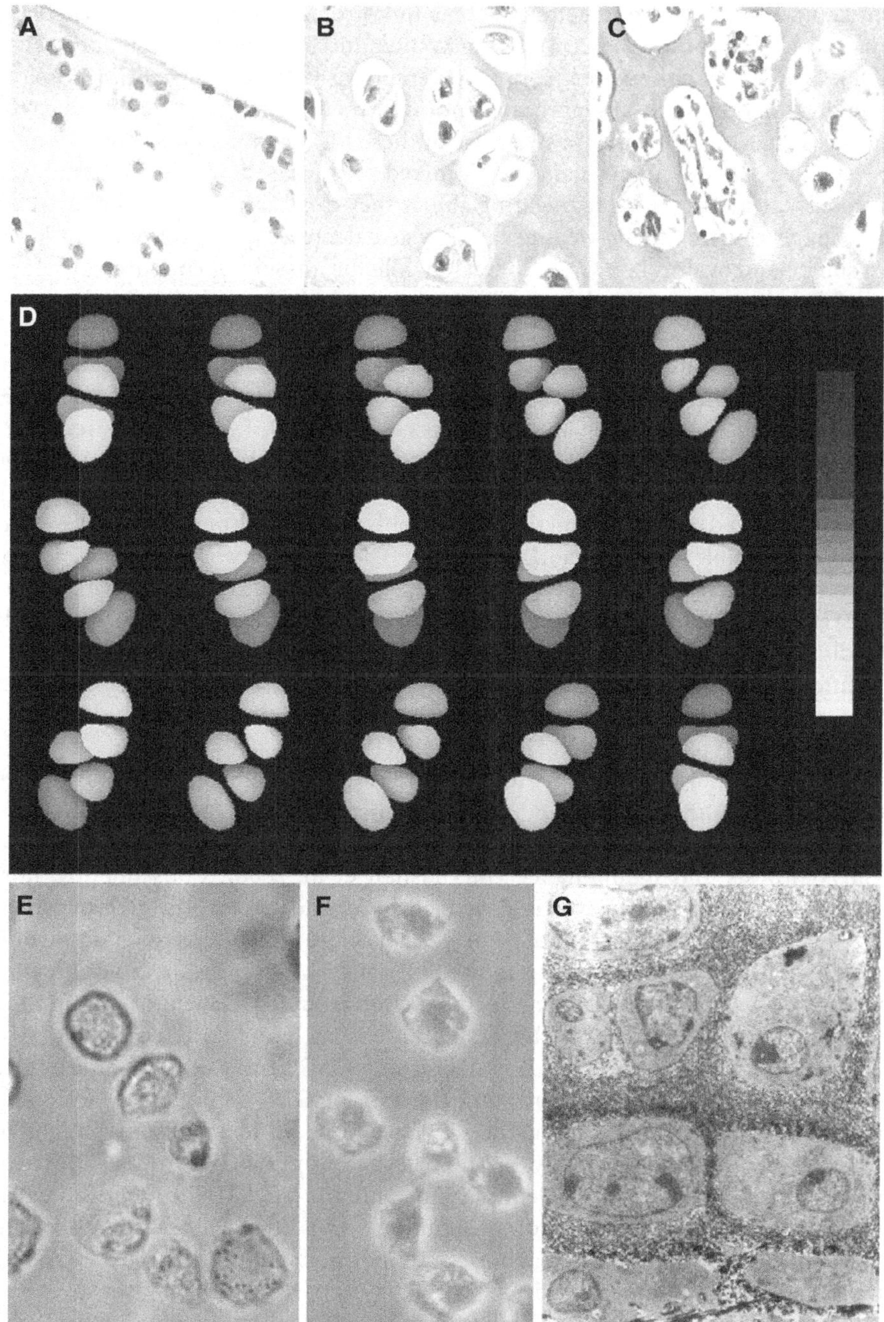

which bind the structure together such as link protein, fibronectin, and cartilage oligomeric matrix protein (COMP) (Buckwalter and Mankin 1998b; Hedbom et al. 1992). COMP is also thought to support interactions in the cartilage ECM, possibly through functioning as an aggrecan-binding protein (Chen et al. 2007). As cartilage ages, the composition and structure of these ECM components change. Cartilage samples from humans and dogs of mixed ages have shown that although the proteoglycans retain their aggregating ability they decrease in size as the cartilage ages, suggesting that their swelling capacity also decreases. In addition, their keratan sulfate and hyaluronan GAGs increase while the amount of chondroitin sulfate decreases, thus suggesting that cartilage loses its ability to withstand and recover from compression as it ages (Bayliss and Ali 1978; Inerot et al. 1978).

Chondrocytes are the only cell type present in articular cartilage and can synthesize all the components of this ECM (Archer and Francis-West 2003; Buckwalter and Mankin 1998b; Kuettner 1992) (Fig. 1). Despite their relatively homogeneous phenotype, chondrocytes vary in size, morphology, and metabolic activity depending on which zone of articular cartilage they are found (Fig. 1A–C). These specialized cells surround themselves with a territorial, pericellular ECM, primarily responsible for mechanotransduction (Guilak 2000; Mobasheri et al. 2002a) and an inter-territorial ECM that makes up the bulk of the load-bearing tissue (Fig. 1D). When isolated and maintained in monolayer culture (Fig. 1E, F), these cells rapidly (after four to five passages) lose their fully differentiated functions and de-differentiate into a fibroblastic phenotype. However, when they are grown in three-dimensional culture, de-differentiated chondrocytes re-express the differentiated phenotype (Benya and Shaffer 1982) (Fig. 1G).

The interaction of chondrocytes with the ECM is mediated by the presence of transmembrane-receptors and proteoglycans, namely, integrins and NG2 on their primary cilia which are involved in the signaling processes required for ECM synthesis and maintenance (McGlashan et al. 2006). Consequently, normal cartilage matrix turnover is governed by chondrocytes, the rate of which is determined by various stimuli. For example, growth factors such as insulin growth factor-II (IGF-II) can stimulate DNA and GAG synthesis in adult cartilage in vitro via their mitogenic effects on chondrocytes (Davenport-Goodall et al. 2004; Henson et al. 1997); whereas cytokines such as interleukin-1β (IL-1β) can cause cartilage degeneration by decreasing GAG synthesis by chondrocytes (Ikebe et al. 1986) and/ or increasing the rate of GAG depletion (MacDonald et al. 1992). Various intracellular and extracellular proteinases are thought to facilitate this depletion, notably the matrix metalloproteinases (MMPs), which are responsible for collagen degradation, and the ADAM-TSs (A disintegrin and metalloproteinase with thrombospondin motifs), which degrade aggrecan (Cawston et al. 1999). Thus the anabolic activities of chondrocytes are in dynamic equilibrium with their catabolic actions, maintaining the homeostasis of the ECM of the articular cartilage. This balance can be affected by many factors including exercise and age. Slight to moderate exercise can have chondroprotective effects via increasing the production of the anti-apoptotic heat shock protein 70 (Hsp70), although this protective effect is reduced as exercise intensity increases (Galois et al. 2004). Similarly, increasing age reduces Hsp90 levels and their functional response to stress which is thought to play a role in the cellular dysfunction observed with ageing (Boehm et al. 2007). Using an Hsp90 inhibitor to simulate reduced Hsp90 levels typical of ageing, it has been shown that chondrocytes are less responsive to insulin-like growth factor-1 (IGF-1) and its anabolic effects such as cell proliferation and collagen type II alpha I (COL2A1) gene upregulation (Boehm et al. 2007). Thus ageing appears to disrupt the normal cellular signaling pathways in articular cartilage.

2.2
Loss of Articular Cartilage Structure and Function in Osteoarthritis

Loss of articular cartilage is the major cause of joint dysfunction and disability in human and animal osteoarthritis (OA) (Buckwalter and Mankin 1998a; Buckwalter et al. 2000). OA is a very common condition in humans and animals; in fact OA is the most common form of arthritis. It affects synovial joints and is often described as 'wear and tear' arthritis. In humans OA may begin in our teens and gradually get worse as we grow older. Alternatively OA can develop in middle age or old age. Many people and animals are predisposed to developing OA. Different prevalences of OA have been cited for different ethnic groups. African–American women appear to have a greater prevalence of knee OA than other groups (Dominick and Baker 2004; Hawker 2004). Additionally, certain breeds of dogs, especially the larger breeds, are genetically predisposed to OA and can develop symptoms at a young age (Liu et al. 2003). Although the following statement might be considered a gross generalization it should be acceptable to propose that OA results from hip dysplasia in German shepherds, Labrador retrievers, and golden retrievers. Also OA may result from elbow dysplasia in Labrador retrievers and golden retrievers and from shoulder dysplasia in Labrador retrievers and Newfoundland breeds. On the other hand, greyhounds, border collies, and whippets are to some degree protected from OA as long as they are protected from traumatic and occupational injuries. Some humans are also protected from OA and it is bewildering how OA can vary from one individual to another. OA usually develops gradually, over time. Several different joints can be affected, but in humans OA is most frequently seen in the hands, knees, hips, feet, and spine (i.e., the intervertebral disc).

Primary OA is the most common noninflammatory arthropathy of both humans and animals as a disorder of movable joints, particularly those that are large and weight bearing. Another term which is commonly used to describe OA is 'osteoarthrosis,' which is virtually identical to 'osteoarthritis' and can be used almost interchangeably. However, 'osteoarthrosis' refers to a condition where the joints are affected by degeneration whereas 'osteoarthritis' implies the same, but the 'itis' at the end adds the additional implication of red, hot, swollen, and painful (inflamed) joints. However, OA is not usually a crippling arthritis in the way that some other forms, such as rheumatoid arthritis, can be. In the majority of cases humans and animals with OA do not have fully inflamed joints, although they may well be painful and, to some extent, deformed. Thus, in reality, 'osteoarthrosis' is probably a more accurate overall description of the condition. However, the majority of individuals that deal with this condition refer to it as 'osteoarthritis,' which is the term most commonly used in the literature.

OA is a chronic degenerate disease with loss of articular cartilage components, particularly type-II collagen and cartilage-specific proteoglycans as the characteristic feature, because of an imbalance between ECM destruction and repair (Todhunter et al. 1996). Although OA chondrocytes have increased expression of both anabolic and catabolic matrix genes (Aigner et al. 2006a), their catabolic ability is thought to outweigh their anabolic capacity resulting in cartilage loss. As OA progresses from mild to severe, there is a decrease in genes coding for transcription of collagen, failure to maintain the proteoglycan matrix, and reduced ability of chondrocytes to regulate apoptosis (Smith et al. 2006). The disease not only affects cartilage but has also been shown to cause damage to the entire joint structure including the subchondral bone, synovium and joint capsule. The exact cause of OA is not yet known and it is often classed as a noninflammatory condition; however, many recent studies have shown that inflammation of the synovium may play an important role in the pathogenesis of the disease (Fiorito et al. 2005). In fact several studies have shown that OA chondrocytes produce inflammatory cytokines and mediators (Cecil et al. 2005; Fernandes et al. 2002).

OA is grossly characterized by aberrant synthesis of articular cartilage matrix, gradual hypocellularity, eventual fragmentation and degradation of articular cartilage, peri-articular new bone formation (osteophytosis), decreased, then increased, subchondral bone density, and variable synovial inflammation (Buckwalter and Mankin 1998a; Buckwalter et al. 2000) (Fig. 2). Furthermore, OA is the clinical phenotype resulting from a number of possible abnormalities of connective tissue function combined with aberrant chondrocyte behavior and an overwhelming of the cartilage's reparative abilities. The pathological changes observed in OA appear to follow cellular and molecular processes involving catabolic and reparative events. In OA, mechanical stress initiates cartilage lesions by altering the chondrocyte–matrix interaction and metabolic responses in chondrocytes (Goldring 2000b). There are initial increases in the amounts of water and proteoglycans associated with the observed transient chondrocyte proliferation of early OA. Proliferating chondrocytes appear in clusters or islands and are accompanied by a change in cellular organization, indicating their undifferentiated nature. In contrast, collagen type X, which is normally produced by terminally differentiated hypertrophic chondrocytes, has been demonstrated surrounding chondrocyte clusters in osteoarthritic cartilage (von der Mark et al. 1992). Chondrocyte proliferation is considered to be an attempt to counteract cartilage degradation, but disease progression and secondary inflammation proves that this is generally unsuccessful. The short-lived hyperplasia (chondrocyte cloning) is followed by hypocellularity and apoptosis (Blanco et al. 1998, 1995; Clegg and Mobasheri 2003; Kim et al. 2003; Mobasheri 2002). Catabolic events responsible for cartilage matrix degradation comprise the release of catabolic cytokines such as IL-1β, IL-6, and tumor necrosis factor (TNF)-α (Goldring 1999, Westacott and Sharif 1996) inducing matrix-degrading enzymes such as MMPs and aggrecanase (ADAM-TS4, ADAM-TS11)

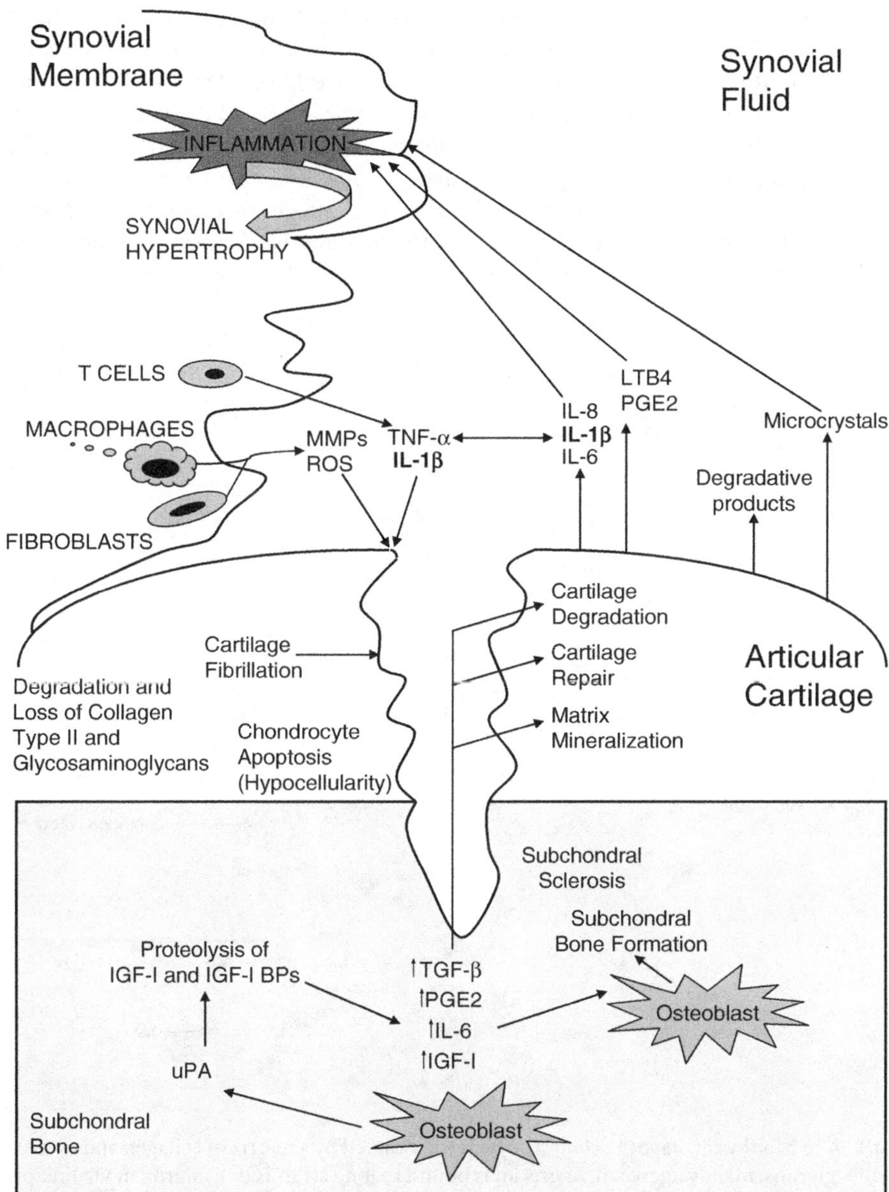

Fig. 2 Summary of the major synovial, chondral and subchondral changes that are thought to occur in OA. This schematic also highlights the actions of various white blood cells and inflammatory mediators in OA. (Modified from Henrotin et al. 2005b)

by chondrocytes and by synoviocytes in early OA (Goldring 1999, 2000a, 2000b; Martel-Pelletier 1998; Westacott and Sharif 1996). Imbalance between MMPs and tissue inhibitors of MMPs occurs, resulting in active MMPs (Fig. 3) and this may be important in cartilage matrix degradation. However, IL-1β may also contribute to the depletion of cartilage matrix by decreasing synthesis of cartilage-specific proteoglycans and collagen type II (Goldring 2000a; Richardson and Dodge 2000; Robbins et al. 2000; Studer et al. 1999). Systemic effects of elevated IL-1β levels include stimulation of glucose transport and metabolism causing hypoglycemia and impairing glucose-induced insulin secretion (del Rey and Besedovsky 1987).

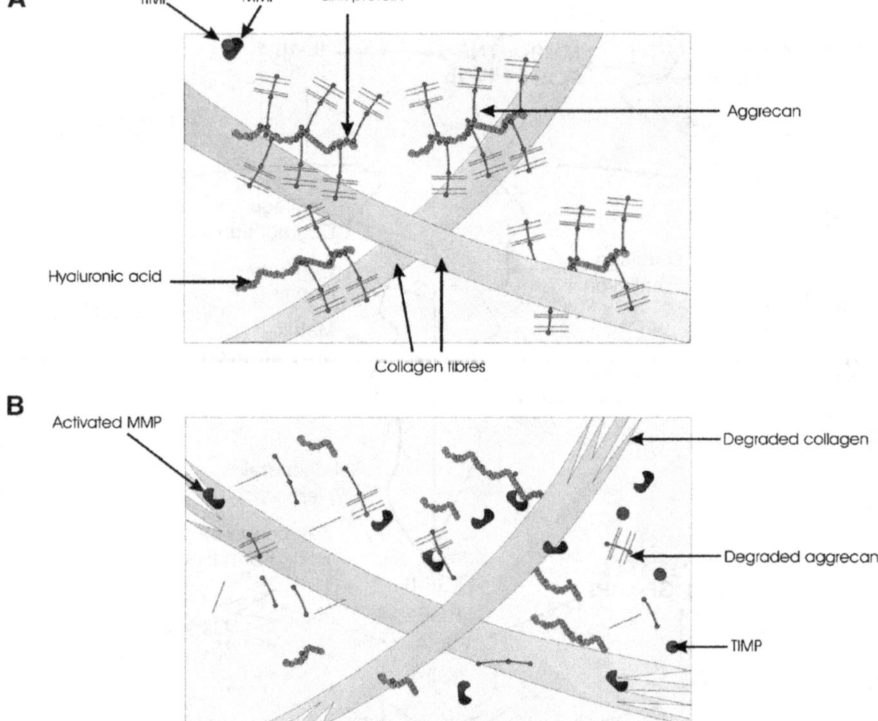

Fig.3 **A** In healthy cartilage the chondrocyte is surrounded by a matrix of collagen and hydrated proteoglycans, mainly aggrecan. Aggrecan is bound to long strands of hyaluronan via link proteins. **B** In OA there is a shift from the normal balance between matrix formation and destruction towards cleavage of matrix components. Aggrecan is primarily severed by aggrecanase and collagens are degraded by collagenases. The enzymes are collectively known as matrix metalloproteinases (MMPs). MMP activity is regulated at multiple levels such as the activation of proenzymes and via suppression by tissue inhibitors of metalloproteinases (TIMPs). TIMPs are tissue inhibitors of MMPs which normally bind to and inhibit these enzymes by forming MMP-TIMP complexes. In OA the ratio of TIMPs to MMPs is altered by the disease process thus increasing MMP activity. [Reproduced from Mobasheri et al. 2002c, with copyright permission of *Histology and Histopathology* (http://www.hh.um.es/) and Jiménez-Godoy, S.A.]

In articular cartilage, the acute effects of IL-1β also involve stimulated glucose uptake and metabolism (Hernvann et al. 1996; Shikhman et al. 2001a). When the matrix is degraded, an inappropriate and inferior repair matrix is synthesized which cannot withstand normal mechanical load. Consequently, cartilage fibrillation and breakdown occurs by the focal formation of vertical, oblique, and tangential clefts into the ECM and is localized preferentially in areas of proteoglycan depletion. Apoptosis is another contributing factor to the loss of articular cartilage in OA: apoptosis increases the cell loss observed in ageing and OA cartilage (Adams and Horton 1998; Blanco et al. 1998; Mobasheri 2002). In addition to deregulated MMP activity and chondrocyte apoptosis, poor diets and malnutrition are also considered to be contributors to the pathogenesis of bone and joint disorders in humans and animals (Kealy et al. 1997; McAlindon and Felson 1997).

3
Does Arthritis Have a Nutritional Etiology?

Despite the recognition that degenerative cartilage disorders like OA and OCD may have nutritional abnormalities at the root of their pathogenesis, the role of nutrition in the etiology of these disorders is poorly studied (Mobasheri et al. 2002c). A huge amount of research effort and funding is focused on nutraceuticals, nutritional supplements, and naturally occurring bioactive components of foods (Goggs et al. 2005; Mobasheri et al. 2002c; Shakibaei et al. 2007a, 2007b, 2005). It is clear that balanced dietary supplementation programs have played a secondary role in the management of joint diseases. Nutritional factors such as glucose and glucose-derived sugars (i.e., glucosamine sulfate and vitamin C) are important for the development, maintenance, repair, and remodeling of cartilage, bone, and other load-bearing connective tissues. In the following sections we review the links between nutrition and joint disease in order to justify the physiological and pathophysiological relevance of our studies on glucose transporters in chondrocytes.

3.1
Nutrition and Osteoarthritis

OA is a multifactorial, polygenic disorder; the pathogenesis of OA involves multiple etiologies, including mechanical, biochemical, environmental, systemic, and genetic factors that contribute to the imbalance in the synthesis and destruction of articular cartilage. Proinflammatory and catabolic cytokines (IL-1β, TNF-α) are involved in disease initiation and progression. Equally, anabolic mediators such as IGF-I, basic fibroblast growth factor (bFGF), connective tissue growth factor (CTGF), and transforming growth factor beta (TGF-β) are responsible for stimulating articular chondrocyte matrix synthetic and mitotic activity, augmenting articular cartilage repair, and inhibiting chondrocyte-mediated matrix catabolism (Haudenschild et al. 2001; Kumar et al. 2001; Pickart and Thaler 1980; Trippel 1995; van den Berg 1999; Yosimichi et al. 2001). However, these are intermediary effectors that can be triggered by a number of other factors.

One important contributing factor to the pathogenesis of OA is diet (Mobasheri et al. 2002c; Schwartz et al. 1981; Wilhelmi 1993b). Nutritional deficiency and imbalance, obesity, and diabetes (resulting in hypoglycemia or hyperglycemia) can result in metabolic and systemic disturbances that in turn increase susceptibility to OA directly due to effects on cartilage. Recent studies suggest that cartilage damage in OA occurs coincident with metabolic dysfunction, nutrient imbalance, and diabetes mellitus (Denko and Malemud 1999; Okma-Keulen and Hopman-Rock 2001; Rosenbloom and Silverstein 1996; Wilhelmi 1993a). Furthermore, high-carbohydrate diets and generalized vitamin deficiency cause metabolic damage to cartilage (Wilhelmi 1993b). Vitamins E, B_2, and C have been shown to exert an inhibitory effect on OA in animals (Wilhelmi 1993b). Chondrotoxic damage may result from food contaminants and fluoroquinolones (Forster et al. 1998; Shakibaei et al. 1996; Stahlmann et al. 1995, 2000). Equally, mineral deficiency (i.e., calcium, magnesium, zinc, selenium, and boron) can provoke skeletal damage in humans and animals.

3.2
Nutrition and Osteochondritis Dissecans

Another joint disorder, which has a dietary component, is osteochondritis dissecans (OCD). In this disorder articular cartilage fragments separate from the articular surface and break off into the joint space (Williams et al. 1998). OCD in food-producing animals is caused by over-nutrition from excess protein and carbohydrate consumption and over supplementation (Slater et al. 1992). OCD in companion animals such as horses has been associated with deficiencies in certain minerals such as copper. It has been shown in cartilage co-cultured with synovial tissue that copper can reverse synovial tissue-stimulated proteoglycan depletion, and levels of 0.01 mM can reduce chondrocyte expression of the catabolic enzymes cathepsin B and cathepsin D (Davies et al. 1996). An early study found that seven out of eight thoroughbred foals which developed osteochondrosis had abnormally low serum copper concentrations (Bridges et al. 1984). Low copper levels may be attributed to a primary deficiency or the secondary effects of other factors that limit copper absorption or metabolism such as sulfates. Water contaminated with sulfates has been associated with the development of osteochondrosis in bison calves through causing a copper deficiency (Woodbury et al. 1999). Similarly, high levels of zinc, such as that found in ponies grazing in close proximity to zinc smelters, are associated with the development of osteochondrosis (Kowalczyk et al. 1986). Consequently researchers investigated the effects of increasing the supplementation of copper to mares in late gestation and to their foals, which had a beneficial effect on reducing the severity of osteochondrosis-type lesions (Knight et al. 1990). However, a recent study has refuted this, finding that neither copper supplementation of the dam nor the liver concentration of copper of the foals had a significant effect on cartilage lesion in sites predisposed to osteochondrosis (Gee et al. 2007). Thus although there is a suspected link, the role between copper and OCD remains unclear.

The overwhelming majority of musculoskeletal problems in companion animals and rapidly growing food-producing animal species are linked with a possible nutritional-related etiology. Therefore, it seems entirely plausible to suggest that a well-balanced, low-fat, and moderate carbohydrate diet together with nutritional supplements, vitamins, and essential minerals will benefit individuals susceptible to degenerative joint disorders. Although nutrition alone will not be responsible for the pathogenesis of OCD or any of the developmental musculoskeletal diseases, it has been suggested that the development and progression of OCD and other orthopedic diseases may be influenced by nutritional management (i.e., by altering food intake and nutrient profile (Richardson and Zentek 1998)). Furthermore, nutraceutical supplements and modulation of nutrient transporters may hold promise for developing future strategies for the prevention, treatment, and cure of OA, OCD, and related joint diseases. This approach may also be of benefit in treating soft connective tissue disorders in humans and animals.

3.3
Nutrient Diffusion in the Extracellular Matrix of Articular Cartilage

Despite more than 30 years of research, the process of nutrient diffusion in articular cartilage is still poorly understood. It is generally accepted that chondrocytes and the matrix are mutually dependent on each other. The matrix is not solely concerned with encapsulating the chondrocyte and protecting it from mechanical damage during joint loading. Many molecules pass through the matrix, including nutrients, substrates for the synthesis of matrix molecules, newly biosynthesized ECM macromolecules, matrix proteases, degraded matrix molecules, metabolic waste products, growth factors, hormones, and cytokines. In some instances these molecules may be stored in the matrix (Buckwalter and Mankin 1998b). The types of molecules that pass through the matrix and the rate at which they can travel depend on the size and charge of the molecule, the composition and organization of the matrix, particularly the concentration, composition, and organization of large proteoglycans. Generally, monovalent and divalent ions, glucose, amino acids, and water freely pass through the ECM (Torzilli et al. 1997, 1998).

How does cartilage obtain its nutrients? This has been a relatively controversial subject in connective tissue research and the answer to this question very much depends on which review article you read. The chondrocyte holds a key position in the maintenance of cartilage matrix and in repair responses during the development of OA. As the only living element of articular cartilage, the chondrocyte produces matrix components using nutrients provided by the surrounding tissues. In the course of its life, the chondrocyte is susceptible to nutritional deficiency and nutritional imbalance (excess quantities of certain nutrients and waste products). Perhaps before discussing the routes for nutrient provision we should clarify exactly which nutrients are needed by chondrocytes. The principal requisites are glucose and other sugars, amino acids, nucleotides, nucleosides, vitamins, and trace minerals, all essential for cartilage matrix synthesis and maintenance.

3.4
Cartilage Canals and Vascular Supply in Subchondral Bone

Around the third fetal month of human development, vascular canals coming from the perichondrium are recognized in the mineralized epiphyseal cartilage. Thus, cartilage end plates contain conduits through which nutrients are provided to developing, immature articular cartilage (Burkus et al. 1993). These cartilage canals are important not only in the nutrition of the chondroepiphysis but also in the initial endochondral osteogenesis of the secondary ossification center (Cole and Wezeman 1985; Yamaguchi et al. 1990). The vascular supply through cartilage canals plays a critical role in developing sufficient biological inertia for the ossifying transition (Ganey et al. 1992). Cartilage canals develop through invasive fibroblast- (of mesenchymal origin) and macrophage-mediated chondrolysis (Chappard et al. 1986; Cole and Wezeman 1987; Skawina et al. 1994). It has been suggested that growth of cartilage canals involves programmed cell death stimulated by thyroid hormone (Delgado-Baeza et al. 1992). Canals are generally believed to close before fully mature cartilage is formed. Furthermore, age-related changes have been observed in the arterioles, capillaries, and venules found in the nutrient canals adjacent to cartilage or intervertebral disc (Bernick and Cailliet 1982). The calcification of the articular cartilage and vascular changes seen in the older vertebrae would be expected to impede the passage of nutrients from the blood to intervertebral disc cartilage.

3.5
Joint Microcirculation

In human, rabbit, and canine articular cartilage there are capillaries running through the subchondral bone in cylindrical channels 20–40 µm wide (Clark 1990). These channels are surrounded by concentric lamellae of bone and only a minority of these channels open into the calcified articular cartilage. Most vascular channels are separated from the cartilage by a defined layer of bone. Hence it has been suggested that vessels within subchondral bone are present primarily to supply the bone with nutrients. The supply of chondrocytes is mediated mainly by the synovial microcirculation system. The microcirculation of the synovial joints is well adapted to its primary function which is to supply nutrients to the avascular cartilage, whose chondrocytes are metabolically active but are relatively large distances from the nearest blood capillary; the nutrients involved may have to diffuse over 1 cm to access chondrocytes in the center of a human knee (Levick 1995). Material exchange is facilitated by a high density of fenestrated capillaries situated very close to the synovial surface with fenestrations preferentially oriented toward the joint cavity. Even so, diffusion alone is too slow to supply central chondrocytes with glucose and other vital nutrients. The problem is partially solved by the synovial microcirculation generating intra-articular fluid (synovial fluid) that transports glucose and other nutrients during dynamic joint loading and physical exercise.

3.6
Diet Influences Cartilage Canal Development

Diet has an important effect on the development of cartilage canals (Woodard et al. 1987b). In an experimental model using piglets, diet has been shown to affect the development of cartilage and its subchondral canals. Piglets fed a reduced protein diet compared to control piglets have been shown to have reduced body weights, reduced longitudinal bone growth, and evidence of early changes associated with OCD (Woodard et al. 1987a). Animals fed a low-protein diet showed disorderly foci of endochondral ossification beneath their articulating cartilages (characterized by an area of chondrocyte necrosis preventing normal cartilage matrix mineralization). These abnormalities have been associated with abnormalities of the cartilage canal vessels, and chondrocyte necrosis was considered to precede degenerative changes in articular cartilage matrix (Woodard et al. 1987b).

4
Metabolic Dysfunction in Arthritis

Healthy bones and joints depend on a normally functioning endocrine system. It is a fact of clinical significance that excess IGF-I and growth hormone (GH) causes major joint pathology (Stavrou and Kleinberg 2001). Endocrine disorders not only affect soft connective tissues but also implicate load-bearing musculoskeletal structures including bone, cartilage, synovium, tendon, and ligament. Damage to soft connective tissue and associated innervation is a hallmark of acromegaly, hypothyroidism, and diabetes mellitus (Liote and Orcel 2000). Acromegaly normally presents with quite severe arthritis involving degeneration of the spine and articular cartilage in peripheral joints (Stavrou and Kleinberg 2001). Severe diabetes mellitus increases the risk of neuroarthropathy as a direct result of infection, neuropathy, and vasculopathy. Pituitary tumors can have manifestations similar to rheumatological disease and may cause connective tissue disorders as a result of overproduction or deficiencies of pituitary hormones (Stavrou and Kleinberg 2001). Excessive GH production by the pituitary gland causes cartilage destruction. GH deficiency on the other hand increases the risk of bone fractures.

4.1
Regulation of Cartilage Turnover by Nutritional and Endocrine Factors

Levels of circulating IGF-I-binding proteins (IGFBPs) are decreased in cases of chronic undernutrition (Underwood 1996). Dietary manipulation can have a significant effect on IGFBPs such as reduction in IGFBP-3, the principal carrier of circulating IGF-I in serum, and under adverse nutritional conditions IGFBP-1 and IGFBP-2 are also affected (Underwood 1996). These changes in IGF-I distribution levels and its clearance will have significant effects on cartilage turnover. Recently, increased IGFBP levels have been found in OA cartilage and chondrocytes (De Ceuninck et al. 2004). This alteration may contribute to the poor anabolic efficacy of IGF-1 in OA cartilage. This study also suggests that a new pharmacological approach that uses a small molecule inhibiting IGF-1/IGFBP interaction could restore or potentiate proteoglycan synthesis in OA chondrocytes thus opening new and exciting possibilities for the treatment of OA and, potentially, of other joint-related diseases.

Leptin is an adipocyte-secreted hormone also known as an adipokine which regulates weight centrally and links nutritional status with neuroendocrine and immune functions. Since its identification and molecular cloning in 1994, leptin's role in regulating immune and inflammatory response has become increasingly evident. A key role has been identified for leptin in OA pathology. It has recently been demonstrated that leptin exhibits, in synergy with other proinflammatory cytokines, detrimental effects on articular cartilage cells by promoting catabolic reactions. Chondrocytes possess the leptin receptor (a product of the obese Ob-R gene), and this receptor is present on chondrocytes in articular human cartilage in situ (Figenschau et al. 2001). Binding of leptin to Ob-R results in phosphorylation of signal transducers and activators of transcription factors (STAT1 and STAT5), and chondrocytes stimulated with leptin exhibit increased proliferation and enhanced synthesis of ECM macromolecules (proteoglycans and collagen). Leptin may also play an important role in endochondral ossification; high levels of leptin expression have been detected in hypertrophic chondrocytes adjacent to capillary blood vessels invading hypertrophic cartilage (Kume et al. 2002). These results suggest that leptin affects cartilage generation directly and that circulating leptin exerts its influence on endochondral ossification by regulating angiogenesis. More recently leptin has also been shown to synergize with IL-1β and interferon gamma (IFN-γ) to induce NO production which will exert catabolic effects on chondrocytes (Otero et al. 2003). These are relatively novel roles for leptin in skeletal growth and development (Otero et al. 2003, 2005, 2006). Collectively, these studies identify a role for nutrition, circulating IGFs, IGFBPs, and leptin in cartilage development and physiological remodeling of cartilage matrix.

4.2
Resistance to IGF-I Implicates Metabolic Dysfunction in OA

Degenerative diseases of load-bearing cartilage and intervertebral disc are generally characterized by disequilibria between ECM repair and degradative processes, with the former not keeping pace, resulting in a loss of proteoglycans, type II collagen, and other ECM components as a result of elevated MMP and aggrecanase activity (Buckwalter and Mankin 1998a; Buckwalter et al. 2000). It is becoming increasingly apparent that a number of degenerative conditions in cartilage occur coincident with endocrine diseases associated with metabolic dysfunction and glucose imbalance (Denko et al. 1994; Denko and Malemud 1999). Independent reports indicate that the reduced growth and repair observed in degenerative cartilage disorders such as OA may be related to an inability of IGF-I to exert its physiological and anabolic effect on chondrocytes. In healthy cartilage IGF-I promotes differentiated cellular functions, which include stimulation of proteoglycan (PG), type-II collagen, and other ECM components. There is an age-related decline in the chondrocyte response to IGF-I and chondrocytes from human OA joints do not respond well to IGF-I stimulation (Dore et al. 1994). This anomaly known as the 'IGF-I-resistant state' implicates insulin/IGF-I signaling and glucose metabolism in the pathogenesis of OA (Kelley et al. 1999).

5
Glucose: An Essential Metabolite and Structural Precursor for Articular Cartilage

Glucose is an important metabolite for all living cells and with other simple sugars and related molecules derived from these sugars it provides sources of readily available energy for cells. Sugars also provide basic carbon skeletons for the biosynthesis of other macromolecules. These include proteins, lipids, nucleic acids, and complex storage polysaccharides (glycogen). Furthermore, hexose sugars are building blocks of glycoproteins such as cartilage-specific proteoglycans. In addition to their role as structural components of the cartilage ECM (Fig. 4), proteoglycans also fulfil adhesive and informational functions.

5.1
Glucose Metabolism in Articular Cartilage

Glucose has been shown to be an important metabolite and structural precursor for cartilage and its regular provision and uptake will have significant consequences for the development and functional integrity of the tissue (Wang et al. 1999). Glucose must diffuse across the synovium and the ECM before reaching chondrocytes (Torzilli et al. 1997). Chondrocytes are highly glycolytic cells and require a regular supply of glucose for optimal ATP production and cell homeostasis (Mobasheri et al. 2005b, 2002c; Mobasheri et al. 2006; Otte 1991). Even modest changes in glucose concentrations in the extracellular microenvironment of chondrocytes [such as those that can occur in poorly treated insulin-dependent diabetes mellitus (IDDM)], could impair IGF-I-mediated anabolic activities and thus promote a variety of joint pathologies (Liote and Orcel 2000; Rosenbloom and Silverstein 1996). Therefore, the steady supply and transport of physiological levels of glucose are critical for chondrocyte viability and matrix synthesis.

5.2
Glycolysis, Glucose Metabolism, and ATP Production in Articular Cartilage

Many different factors affect the delivery of glucose to peripheral tissues and its removal from the blood. These factors include dietary carbohydrate and fat, the circulating free fatty acid concentration, exercise, hypoxia, and the actions of

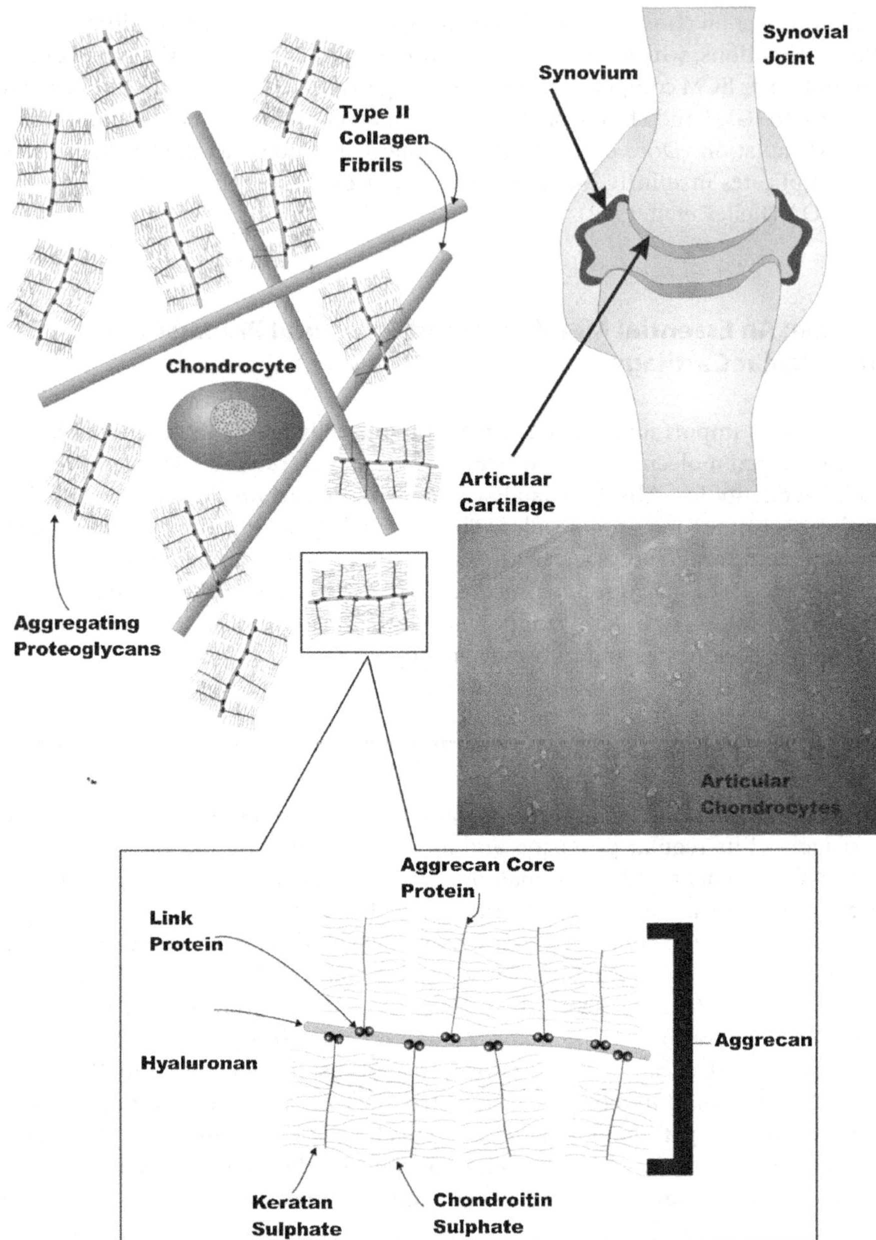

Fig. 4 Structure and composition of the extracellular matrix of articular cartilage. This schematic model of the extracellular matrix represents an articular chondrocyte surrounded by a matrix of type II collagen and proteoglycans. Diffusion of large substances is impeded by the ECM but cations, gases and solutes such as sugars and amino acids freely diffuse and move within the matrix

several metabolic hormones (Casey 2003). The postprandial blood glucose concentration is about 4–5.5 mmol/l or 70–100 mg/100 ml, but arterial concentrations vary throughout the day from about 3.5 mmol/l after exercise to 9 mmol/l following a meal (Casey 2003). For the purposes of this review it is the plasma fraction of blood glucose that is most pertinent to our focus because the plasma glucose concentration should be fairly similar to that found in synovial fluid in normal joints.

Typical mammalian cells generate energy in the form of ATP by anaerobic or aerobic metabolism of glucose. In articular cartilage, the chondrocyte obtains glucose and oxygen by diffusion from synovial fluid. The importance of chondrocyte glucose transport and metabolism for the synthesis of a normal, mechanically competent ECM is emphasized in Fig. 5. The concentrations of glucose and oxygen within cartilage matrix gradually diminish with increasing proximity to the calcified cartilage layer and this is particularly relevant in mature cartilage. The gradients for glucose and oxygen depend on chondrocyte density and metabolic consumption of the cells. The gradient of the partial pressure of oxygen (PO_2) provides the conditions for a negative Pasteur effect in chondrocytes (Lee and Urban 1997). Anoxia severely inhibits glucose uptake and lactate production in cartilage; the decrease in lactate formation correlates well with decreased glucose uptake by chondrocytes and this reduction in the rate of glycolysis in anoxic conditions is seen as evidence for a 'negative Pasteur effect' in cartilage (Lee and Urban 1997). Conversely, in the intervertebral disc, anoxia and addition of glycolysis inhibitors result in a progressive positive Pasteur effect suggesting that unlike articular cartilage, a large proportion of the intervertebral disc's energy derives from oxidative phosphorylation (Ishihara and Urban 1999). Manometric (Warburg technique) studies of porcine femoral head cartilage clearly show a close correlation between oxygen consumption and the concentration and consumption of glucose (the Crabtree effect; Otte 1991). Excess glucose is channelled into synthetic and storage processes consistent with the physiological (ECM synthesis) role of chondrocytes in cartilage matrix (Otte 1991). Under

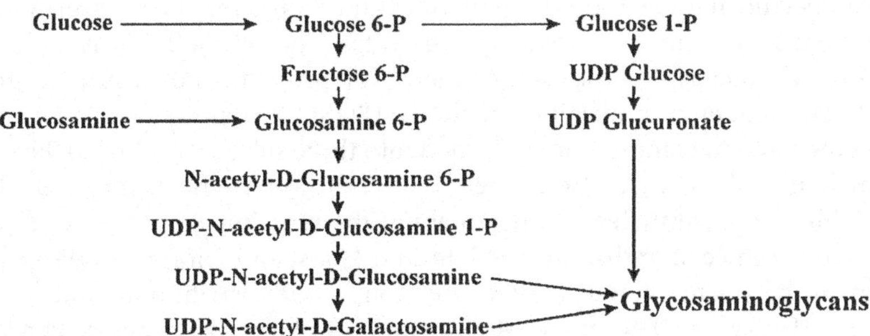

Fig. 5 The biochemical pathway for the synthesis of glycosaminoglycans from glucose and glucosamine as primary substrates

physiological conditions, glucose in the synovium reduces the consumption of oxygen in the well-nourished superficial layer of articular cartilage, thus allowing an oxidative compensation for the diminishing glycolysis rates in the deep zone. Thus the Crabtree effect has an important regulatory role in the basic metabolism of articular cartilage.

6
Mammalian Sugar Transporter Families: GLUT and SGLT

Sugar transport across the plasma membrane of mammalian cells is mediated by members of the GLUT/SLC2A family of facilitative sugar transporters (Joost and Thorens 2001; Mueckler 1994; Seatter and Gould 1999; Uldry and Thorens 2004) and the SGLT/SLC5A family of Na+-dependent sugar transporters (Wood and Trayhurn 2003). These proteins belong to a larger superfamily of proteins known as the major facilitator superfamily (MFS) or uniporter-symporter-antiporter family (Saier et al. 1999a) The MFS family is one of the two largest families of membrane transporters in nature and accounts for nearly half of the solute transporters encoded within the genomes of microorganisms (bacteria, yeasts) and higher organisms such as plants and animals. The MFS was originally thought to function primarily in the uptake of sugars, but more detailed studies of members of this family have revealed that drug efflux systems, Krebs cycle metabolites, organophosphate : phosphate exchangers, oligosaccharide : H1 symport permeases, and bacterial aromatic acid permeases are also members of the MFS superfamily. These observations led to the probability that the MFS is far more widespread in nature and far more diverse in function than had been thought previously (Pao et al. 1998; Saier et al. 1999a, 1999b, 1998). Thus far 17 subgroups of the MFS have been identified. The human genome project has identified 14 members of the GLUT/SLC2A family which have been cloned in humans (Wood and Trayhurn 2003; Wu and Freeze 2002) (Fig. 6, Table 1). GLUT proteins are characterized by the presence of 12 membrane spanning helices and several conserved sequence motifs (Joost and Thorens 2001). The GLUT/SLC2A proteins are expressed in a tissue- and cell-specific manner and exhibit distinct kinetic and regulatory properties that reflect their functional and tissue-specific roles. The full definitions and unique functional characteristics for each of the GLUT protein isoforms are outlined in Table 1. The sequence similarities of the GLUT/SLC2A family members are now well defined and the family is now divided into three subclasses (Fig. 6). Five of the mammalian facilitated glucose carriers (GLUTs 1–5) have been very well characterized, but significantly less is known about the remaining nine glucose carriers (GLUTs 6–14) since their discovery in late 2001 (Joost and Thorens 2001) and much remains to be learned about their expression, tissue distribution, and transport functions (Uldry and Thorens 2004). Of these 14 glucose transporter genes, nine are known to be expressed at the mRNA level in chondrocytes (Richardson et al. 2003) (see subsequent sections).

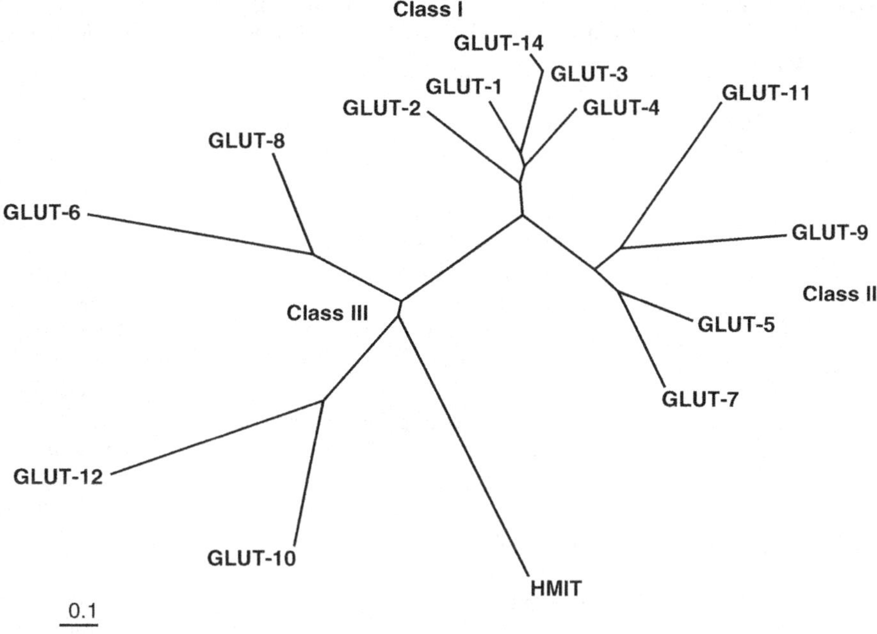

Fig. 6 Members of the extended GLUT/SLC2A family. The radial phylogram was derived from a multiple sequence alignment of the 14 known members of the human GLUT/SLC2A family. The tree was constructed using neighbor-joining analysis of a distance matrix generated with PHYLIP software. The *scale bar* represents 0.1 substitutions per amino acid position. The family is divided into three classes of GLUT proteins; class I includes the well studied GLUTs1–4 and GLUT14; class II includes the fructose transporter GLUT5, GLUT7, GLUT9 and GLUT11; class III includes GLUT6, GLUT8, GLUT10, GLUT12 and the H+-coupled myo-inositol transporter, HMIT. (Reproduced from Wood and Trayhurn 2003, with kind permission of the authors)

6.1
Physiological Roles of GLUTs 1–5

Glucose transporters are integral membrane glycoproteins involved in transporting glucose into most cells. It is generally accepted that GLUT1, GLUT3, and GLUT4 are high-affinity transporters whereas GLUT2 is a low-affinity transporter isoform; GLUT5 is primarily a fructose carrier (Thorens 1996). High-affinity transporters are found in many metabolically active tissues, but their expression is higher in cells with a high glycolytic activity (i.e., hepatocytes, absorptive intestine epithelial cells, and proximal tubule cells of the kidney nephron) (Tal et al. 1990; Thorens et al. 1990).

Table 1 Members of the GLUT/SLC2A family of facilitative glucose/polyol transporters (extended and modified from Airley and Mobasheri 2007, Goggs et al. 2005, Mobasheri et al. 2002c)

Isoform	Previous name	Class	Tissue localization	Insulin sensitive	Substrate
GLUT1	-	I	Ubiquitous: erythrocytes, brain, cartilage. Over-expressed in many tumors and tumor derived cell lines.	No	glucose
GLUT2	-	I	Liver, pancreas, intestine, kidney	No	glucose (low affinity); fructose
GLUT3	-	I	Brain, cartilage	No	glucose (high affinity)
GLUT4	-	I	Heart, muscle, adipose tissue, brain	Yes	glucose (high affinity)
GLUT5	-	II	Intestine, testis, kidney	No	fructose; glucose (very low affinity)
GLUT6	GLUT 9	III	Brain, spleen, leukocytes	No	glucose
GLUT7	-	II	Intestine	No	glucose and fructose
GLUT8	GLUT X1	III	Testes, brain and other tissues	Yes (in the blastocyst)	glucose
GLUT9	GLUT X	II	Liver, kidney, placenta, cartilage	n.d.	n.d.
GLUT10	-	III	Liver, pancreas	No	glucose
GLUT11	GLUT 10	II	Heart, muscle	No	glucose (low affinity); fructose (long form)
GLUT12	GLUT 8	III	Heart, prostate, muscle, small intestine, adipose tissue, cartilage	Yes	n.d.
HMIT (GLUT13)	-	III	Brain	n.d.	H^+/myo-inositol
GLUT14	-	I	Testis	n.d.	n.d.

GLUT11 occurs in two splice variants: a short form (low affinity glucose transport) and a long form (which may be a fructose transporter). The presence of each transporter in skeletal muscle and adipose tissue is shown since these are the major sites of insulin-stimulated glucose uptake; n.d., not determined

GLUT1 originally cloned in human HepG2 hepatoma cells (Mueckler et al. 1985) is a major glucose transporter in the mammalian blood–brain barrier. It is present at high levels in human and primate erythrocytes and in brain endothelial cells.

According to AceView (http://www.ncbi.nlm.nih.gov/IEB/Research/Acembly/index.html) of the National Centre for Biotechnology Information (NCBI http://www.ncbi.nlm.nih.gov/), the GLUT1 gene is expressed at a very high level (15.2 times the average gene in the August 2005 revision of the human genome). The sequence of this gene is defined by 2,178 GenBank accessions, some from placenta (seen 31 times), liver (26), hepatocellular carcinoma, cell line (24), placenta cot 25-normalized (23), placenta normal (20), tongue, tumor tissue (14), embryonic stem cells (13), and 80 other tissues. The regulation of GLUT1 expression is highly complex. There is evidence for alternative mRNA variants and different modes of regulation. The gene contains 21 different GT-AG introns. Transcription produces 11 different mRNAs, eight alternatively spliced variants, and three unspliced forms. There are three probable alternative promoters, five nonoverlapping alternative last exons, and six validated alternative polyadenylation sites. The mRNAs encoding GLUT1 appear to differ by truncation of the 5′ end, truncation of the 3′ end, presence or absence of nine cassette exons, overlapping exons with different boundaries, alternative splicing or retention of three introns. Functionally, the GLUT1 gene has been extensively studied for association to numerous human diseases including: adenocarcinoma; carbohydrate metabolism, inborn errors of metabolism; colorectal neoplasms; diabetes mellitus; diabetic angiopathies; diabetic nephropathies; diabetic neuropathies; epilepsy; glucose transport defects across the blood–brain barrier; congestive heart failure; ovarian neoplasms. GLUT1 has been proposed to participate in various pathways (adipocytokine signaling pathway, vitamin C transport in the brain) and processes (glucose transport, carbohydrate transport). Proteins are expected to have molecular functions (glucose transporter activity and protein binding) and to localize in various compartments (membrane fraction, cytoplasm, endoplasmic reticulum membrane, mitochondrion, plasma membrane, and six other subcellular compartments). The major facilitator superfamily MFS_1 motif (http://www.sanger.ac.uk/cgi-bin/Pfam/getacc?PF07690) is seen in the product of this gene and other glucose transporters and other members of the sugar (and other substrate) transporter gene superfamilies. GLUT1 is expressed in many other human tissues (Fig. 7) including articular cartilage. GLUT1 is abundantly expressed in the brain (Flier et al. 1987a), erythrocytes (Mueckler et al. 1985), and the liver, but is present in significantly lower quantities in cardiac and skeletal muscle which express other glucose transporters including GLUT3 (Guillet-Deniau et al. 1994; Hocquette and Abe 2000; Shepherd et al. 1992) and GLUT4 (Charron et al. 1989; James et al. 1988). Prolonged exposure to hypoxia results in enhanced transcription of the GLUT1 glucose transporter gene in the brain and in many cells lines (Badr et al. 1999; Behrooz and Ismail-Beigi 1997; Bruckner et al. 1999; Hamrahian et al. 1999; Ouiddir et al. 1999; Vannucci et al. 1998, 1996; Ybarra et al. 1997; Zhang et al. 1999). GLUT1 is also significantly upregulated in many tumors (for a review see Airley and Mobasheri 2007). Elevated levels of the GLUT1 glucose transporter are induced by *ras* or *src* oncogenes (Flier et al. 1987b) and a role for this glucose transporter has been proposed in oncogenic transformation and tumor development (Burstein et al. 1998; Mellanen et al. 1994; Nagamatsu

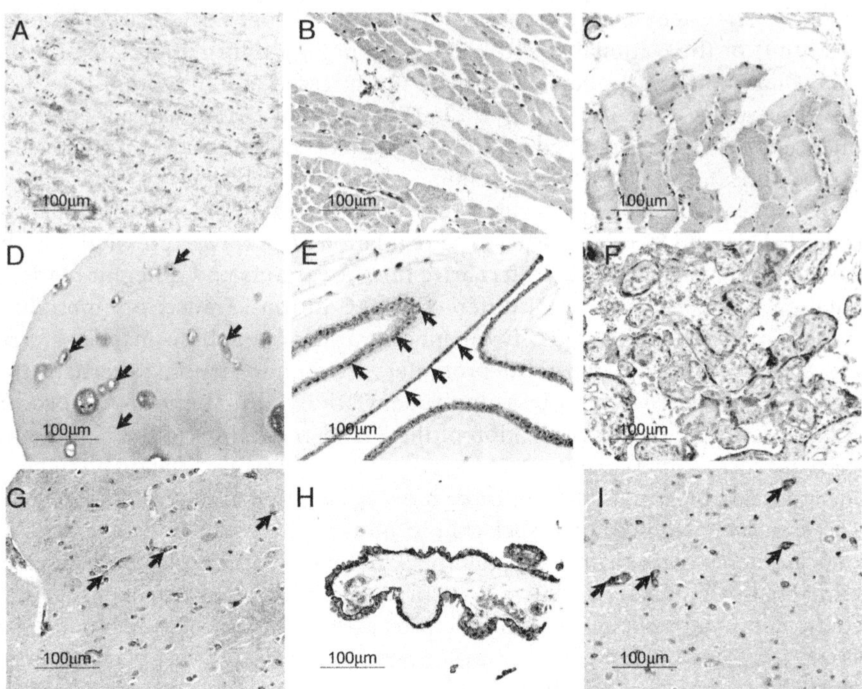

Fig. 7 Differential distribution of the GLUT1 glucose transporter in human tissues. GLUT1 is not detectable in aortic smooth muscle (**A**) but is moderately expressed in cardiac muscle (**B**) and skeletal muscle (**C**). GLUT1 is present in articular chondrocytes (**D**), amniotic membranes (**E**) (Wolf and Desoye 1993), placenta (**F**) (Arnott et al. 1994) and in microvessels (Takakura et al. 1991; Virgintino et al. 1997) in various regions of the central nervous system including cerebral cortex (**G**), choroid plexus (**H**) and hippocampus (**I**). Expression in the choroid plexus was predominantly basolateral (Kurosaki et al. 1995). *Arrows* indicate sites of GLUT1 expression. Normal human Tissue MicroArrays (TMAs) were obtained from the Cooperative Human Tissue Network (CHTN) of The National Cancer Institute (NCI), the National Institutes of Health, Bethesda, MD, USA (http://faculty.virginia.edu/chtn-tma/home.html) and used for GLUT1 immunohistochemistry. The primary antibody used for GLUT1 immunostaining was a rabbit polyclonal antibody (gift of Dr. S. Baldwin, University of Leeds, UK) to an epitope in the C-terminus of the GLUT1 protein (residues 477–492 of human GLUT1). This antibody recognizes GLUT1 across a diverse range of mammalian species including human, canine, equine and ovine. Immunohistochemistry was carried out using a DakoCytomation EnVision+ Dual Link System-HRP (DAB+) kit (Code K4065; Ely, Cambridgeshire, UK) according to the manufacturer's instructions. *Bars* represent 100 µm

et al. 1993; Semenza et al. 2001; Younes et al. 1996). One of the reasons for this apparent increase in GLUT1 in cancer has been proposed to be the hypoxia responsiveness of GLUT1, which may confer a metabolic advantage to tumor cells (Airley and Mobasheri 2007).

GLUT2 is an integral plasma membrane glycoprotein of the liver, pancreatic islet beta cells, intestine, and kidney epithelium. It mediates facilitated bidirectional glucose transport. Because of its low affinity for glucose, it has been suggested as a glucose sensor. GLUT2 protein is present in hepatocytes in the liver, pancreatic beta cells, intestinal epithelial cells, and renal epithelial cells (Fukumoto et al. 1988). According to AceView, the GLUT2 gene is well expressed. The sequence of the GLUT2 gene is defined by 91 GenBank accessions from 85 cDNA clones, some from liver (seen over 30 times), liver and spleen (20), kidney (10), corresponding noncancerous liver tissue (6), liver, tumor tissue (5), hepatocellular carcinoma (3), two pooled tumors (clear cell type) (once), and eight other tissues. Functionally, the GLUT2 gene has been studied for its association to the following human diseases: inborn errors of carbohydrate metabolism, noninsulin-dependent diabetes mellitus, diabetes mellitus, type 2; Fanconi syndrome; Fanconi-Bickel syndrome; glucose intolerance; maturity-onset diabetes of the young and type II diabetes mellitus. GLUT2 has been shown to participate in the following processes and pathways: glucose transport and carbohydrate metabolism. GLUT2 effectively mediates facilitated bidirectional glucose transport. Because of its low affinity for glucose, and its high expression in tissues carrying large glucose fluxes (such as pancreatic β cells, intestine, kidney, and liver), it has been suggested as a glucose sensor (Thorens 1996). GLUT2 has also been proposed to function as a glucose sensor in the brain where it is involved in maintaining glucose homeostasis, and in cells where glucose-sensing is necessary (i.e., hypothalamic neurons) (Waeber et al. 1995). GLUT2 has been implicated in diabetes mellitus (noninsulin-dependent) (Mueckler et al. 1994) and in Fanconi-Bickel syndrome (Santer et al. 2002) (http://www. ncbi.nlm.nih.gov/IEB/Research/Acembly/av.cgi?exdb=AceView&db=35g&term=G LUT2). Indeed in many experimental models of diabetes, GLUT2 gene expression is decreased in pancreatic β cells, which could lead to a loss of glucose-induced insulin secretion and thus to significantly attenuated glucose-sensing capacity.

GLUT3 is encoded by SLC2A3 the third member of the solute carrier family 2 (facilitated glucose transporter), member 3 (Bell et al. 1990; Kayano et al. 1990, 1988). According to AceView, the GLUT3 gene is expressed at very high levels (7.3 times the average gene in the April 2007 release of the Human Genome Project, the most recent release to date). The sequence of the GLUT3 gene is defined by 3,556 GenBank accessions. Many of these are derived from embryonic tissues and stem cells. Other sequences are derived from retinoic acid and mitogen-treated hes cell line H7 (seen 34 times), embryonic stem cell lines (26), liver (24), brain (23), embryoid bodies derived from H1, H7, and H9 cells (23), small intestine (23), duodenal adenocarcinoma cell line (22) and 84 other tissues. GLUT3 gene transcription produces 15 different mRNAs, 11 alternatively spliced variants, and four unspliced forms. There are four probable alternative promoters, four nonoverlapping alternative last exons, and nine validated alternative polyadenylation sites. The mRNAs encoding GLUT3 appear to differ by truncation of the 5′ end, truncation of the 3′ end, presence or absence of a cassette exon, overlapping exons with different boundaries, alternative splicing, or retention of five introns.

Functionally, the GLUT3 gene has been proposed to participate in glucose transport, vitamin C transport in the brain, and various other carbohydrate-dependent metabolic processes.

Early mRNA and protein studies on GLUT3 indicated that transcripts encoding GLUT3 are abundant in the brain and are also present in most other tissues (Arnott et al. 1994; Burant and Davidson 1994; Haber et al. 1993; Maher and Simpson 1994; Maher et al. 1994; Shepherd et al. 1992; Vannucci 1994), although their relative abundance varies. Studies of GLUT3 expression and regulation in the brain suggest that this protein is hypoxia-responsive-like GLUT1 (Badr et al. 1999; Bruckner et al. 1999; Vannucci and Vannucci 2000; Zhang et al. 1999).

The GLUT4 gene is a well studied member of the solute carrier family 2 (facilitated glucose transporter) and encodes a protein that functions as an insulin-regulated facilitative glucose transporter. According to AceView, the GLUT4 gene is well expressed. The sequence of the GLUT4 gene is defined by 53 GenBank accessions, some from heart (seen four times), neuroblastoma cot 10-normalized (4), pectoral muscle (after mastectomy) (4), two pooled tumors (renal clear cell type) (3), kidney (3), prostate (3), bone (2), and 18 other tissues. GLUT4 gene transcription produces five different mRNAs, four alternatively spliced variants, and one unspliced form. There are two nonoverlapping alternative last exons and three validated alternative polyadenylation sites. The mRNAs appear to differ by the presence or absence of a cassette exon, overlapping exons with different boundaries. Mutations in this gene have been associated with noninsulin-dependent diabetes mellitus (NIDDM). Functionally, the GLUT4 gene has been tested for association to the following human diseases: diabetes mellitus, (noninsulin-dependent), diabetes mellitus (type 1 and type 2), congestive heart failure, insulin resistance; diabetes mellitus (type II). The GLUT4 gene has also been associated with participation in metabolic pathways (adipocytokine signaling pathways, growth hormone signaling pathway, insulin signaling pathways) in addition to glucose transport, carbohydrate metabolism, and glucose homeostasis). Various putative intracellular proteins have been proposed to interact with GLUT4. These include AKT2, DAXX, EHD2, SORT1, and UBE2I.

The GLUT4 protein is the major insulin-responsive glucose transporter expressed in cardiac muscle, skeletal muscle, and adipose tissue (Bell et al. 1989; Birnbaum 1989; Fukumoto et al. 1989). Glucose uptake by these tissues is acutely regulated by insulin, which stimulates facilitative glucose transport, at least in part, by promoting the translocation of transporters from an intracellular pool to the plasma membrane. Studies by Garvey and co-workers (Garvey et al. 1998) suggested that insulin alters the subcellular localization of GLUT4 vesicles in muscle, and that this effect is impaired equally in insulin-resistant subjects with and without diabetes. The translocation defect was found to be associated with abnormal accumulation of GLUT4 in a dense membrane compartment demonstrable in basal muscle. A similar pattern has been observed in insulin-resistant human adipocytes leading to the conclusion that insulin resistance involves a defect in GLUT4 traffic and targeting (i.e., in the absence of insulin, this integral membrane protein is sequestered within the cells of muscle and adipose tissue; within minutes of insulin stimulation, the protein moves to the cell surface and begins to transport glucose across the cell membrane).

GLUT5 is the product of the SLC2A5 gene. According to AceView, this gene is expressed at very high level, 4.5 times the average gene in this release. The sequence of the GLUT5 gene is defined by 318 GenBank accessions from 301 cDNA clones, some from testis (seen 46 times), kidney (32), brain (21), germinal center B cell (20), prostate (20), lymph (12), Burkitt's lymphoma (10), and 62 other tissues. GLUT5 transcription produces 17 different mRNAs, 16 alternatively spliced variants, and one unspliced form. There are seven probable alternative promoters, four nonoverlapping alternative last exons, and two validated alternative polyadenylation sites. The mRNAs appear to differ by truncation of the 5′ end, truncation of the 3′ end, presence or absence of 17 cassette exons, overlapping exons with different boundaries, alternative splicing or retention of one intron. Functionally, the GLUT5 gene has been studied for its role in hexose and pentose transport and association to diabetes mellitus, type 2. The GLUT5 protein is the fructose transporter expressed on the brush border membrane of small intestinal enterocytes and spermatozoa (Burant et al. 1992; Davidson et al. 1992). Its localization on the luminal surface of mature absorptive epithelial cells implied that GLUT5 participates in the uptake of dietary sugars (Davidson et al. 1992), a finding that has been confirmed by subsequent functional studies (Mahraoui et al. 1994).

As we have stated earlier, the expression of many sugar transporters is influenced by the availability of the metabolic substrate(s) they carry. These are classic examples of gene regulation by the substrate(s). Also, GLUTs are regulated by endocrine factors; as an adaptive response to variations in metabolic conditions, the expression of the GLUT1–5 transporters is regulated by glucose and different hormones (Thorens 1996).

6.2
Physiological Roles of GLUTs 6–14

As we have described in the previous section, hexose transport into mammalian cells is also catalyzed by recently identified members of the GLUT family of membrane proteins, including GLUT6-GLUT14 (Joost et al. 2002; Joost and Thorens 2001; Wood and Trayhurn 2003). The identification of additional members of the human/mammalian glucose transporter family created a confusing nomenclature for a few years in the late 1990s before a systematic nomenclature was introduced by a consortium of glucose transporter researchers in 2002 (Joost et al. 2002). Significantly less information is available about the physiological roles of the remaining nine glucose transporters (GLUTs 6–14). We would like to direct the readers to several recent reviews (Joost and Thorens 2001; Scheepers et al. 2004; Uldry and Thorens 2004; Wood and Trayhurn 2003).

GLUT6 [SLC2A6] is a protein that contains 12 transmembrane domains and a number of critical conserved residues (Joost and Thorens 2001; Uldry and Thorens

2004). According to AceView, the GLUT6 gene is expressed at high levels (2.6 times the average gene in the April 2007 release). The sequence of this gene is defined by 183 GenBank accessions from 176 cDNA clones. Many of the cDNA clones are from brain (seen 26 times), lung (15), hippocampus (9), spleen (9), cerebellum (8), pancreas (7), neuroblastoma (6), and 43 other tissues. GLUT6 has been proposed to function as a glucose transporter as there are two articles specifically referring to this gene in PubMed. Functionally, sthe gene has been proposed to participate in carbohydrate transport. Recent studies have shown that GLUT6 is principally detected in testis germinal cells (Godoy et al. 2006).

GLUT7 [SLC2A7] was recently cloned from a human intestinal cDNA library by using a PCR-based strategy (GenBank accession no. AY571960) (Li et al. 2004). The encoded protein consists of 524 amino acid residues and shares 68% similarity and 53% identity with GLUT5, its most closely related isoform in class II of the GLUT/SLC2A gene family. When GLUT7 was expressed in *Xenopus laevis* oocytes, it showed high-affinity transport for glucose (K_m=0.3 mM) and fructose (IC_{50}=0.060 mM). The fructose transporting function of GLUT7 has also been investigated by mutagenesis (Manolescu et al. 2005). GLUT7 did not transport galactose, 2-deoxy-D-glucose, and xylose. Also, glucose transport by GLUT7 was not inhibited by 200 µM phloretin or 100 µM cytochalasin B suggesting that this transporter is insensitive to these inhibitors. Northern blot analysis indicated that the mRNA for GLUT7 is present in the human small intestine, colon, testis, and in the prostate gland. Western blotting and immunohistochemistry of rat tissues with an antibody raised against the predicted COOH-terminal sequence confirmed expression of the protein in the small intestine and indicated that the transporter is predominantly expressed in the brush-border membrane of enterocytes. The unusual substrate specificity and close sequence identity of GLUT7 and GLUT5 suggest that GLUT7 represents an intermediate between class II GLUTs and the class I member GLUT2. Regarding expression in other tissues, according to AceView, the GLUT7 gene is expressed at very low level, only 4.5% of the average gene in this release. The sequence of this gene is defined by a small number of GenBank accessions, some from blood vessels.

According to AceView GLUT8 [SLC2A8] is expressed at high levels in the human body, 2.2 times the average gene in the most recent release. The sequence of this gene is defined by 152 GenBank accessions from 144 cDNA clones, some from brain (seen 25 times), testis (17), hippocampus (8), lung (8), ovary (7), whole brain (7), glioblastoma (pooled) (6), and 70 from other tissues. GLUT8 transcription produces 14 different mRNAs, 13 alternatively spliced variants, and one unspliced form. There are five probable alternative promoters, two nonoverlapping alternative last exons, and seven validated alternative polyadenylation sites. The mRNAs appear to differ by truncation of the 5′ end, truncation of the 3′ end, presence or absence of seven cassette exons, overlapping exons with different boundaries, alternative splicing or retention of two introns. Functionally, the GLUT8 gene has been proposed to participate in the following processes: carbohydrate metabolism, glucose transport, insulin receptor signaling, and response to hypoxia. Early studies suggest that GLUT8 is a glucose transporter responsible for insulin-stimulated glucose uptake in the blastocyst (Carayannopoulos et al. 2000).

For a detailed discussion of the physiological functions of GLUTs and more information about GLUT9, GLUT10, GLUT11, GLUT12, HMIT, and GLUT14, we would like to refer the readers to the AceView site (http://www.ncbi.nlm.nih.gov/IEB/Research/Acembly/) and a recent review by Wood and Trayhurn (Wood and Trayhurn 2003).

7
Molecular Diversity of Facilitative Glucose Transporters in Articular Chondrocytes

In order to appreciate the physiological basis for the molecular diversity of facilitative glucose transporters in chondrocytes we need to re-examine the importance of glucose as a nutrient and structural precursor. Glucose is an important nutrient in fully developed articular cartilage due to the poor vascularization and highly glycolytic nature of the tissue, a situation that is further exacerbated by low oxygen tensions and ongoing anaerobic glycolysis by chondrocytes (Mobasheri et al. 2002c; Otte 1991; Rajpurohit et al. 2002). Therefore, even modest changes in glucose concentrations in the extracellular microenvironment of chondrocytes could impair anabolic and catabolic activities (Mobasheri et al. 2002c; Shikhman et al. 2001a). Fully developed adult chondrocytes express mRNA for multiple isoforms of the GLUT/SLC2A family of glucose transporters including GLUT1, GLUT3, GLUT5, GLUT6, GLUT8, GLUT9, GLUT10, GLUT11, and GLUT12 (Mobasheri et al. 2002b, 2002c; Richardson et al. 2003; Shikhman et al. 2001a) (Fig. 8). The reason for such GLUT isoform diversity in chondrocytes has not yet been satisfactorily explained but several hypotheses have been put forward: GLUT isoform diversity in chondrocytes suggests that the transmembrane uptake of glucose, fructose, and other related hexose sugars is highly specialized and requires several proteins with the capacity to transport structurally different sugars. The observed diversity of GLUT proteins in chondrocytes may possibly reflect a cartilage-specific requirement for 'fast' (i.e., GLUT3) and baseline (GLUT1) glucose transporters that operate more efficiently at low substrate concentrations under physiological conditions (Mobasheri et al. 2002c; Richardson et al. 2003). The presence of GLUT1 in chondrocytes has also been linked to the acute requirement of these cells for glycolytic energy metabolism under the low oxygen tension conditions that are prevalent in avascular load-bearing articular cartilage and intervertebral disc (Pfander et al. 2003; Rajpurohit et al. 2002; Schipani et al. 2001). GLUT1 has also been shown to be a cytokine inducible glucose transporter in cartilage since it is induced by catabolic, proinflammatory cytokines (Phillips et al. 2005a; Richardson et al. 2003; Shikhman et al. 2004, 2001a) (Fig. 9).

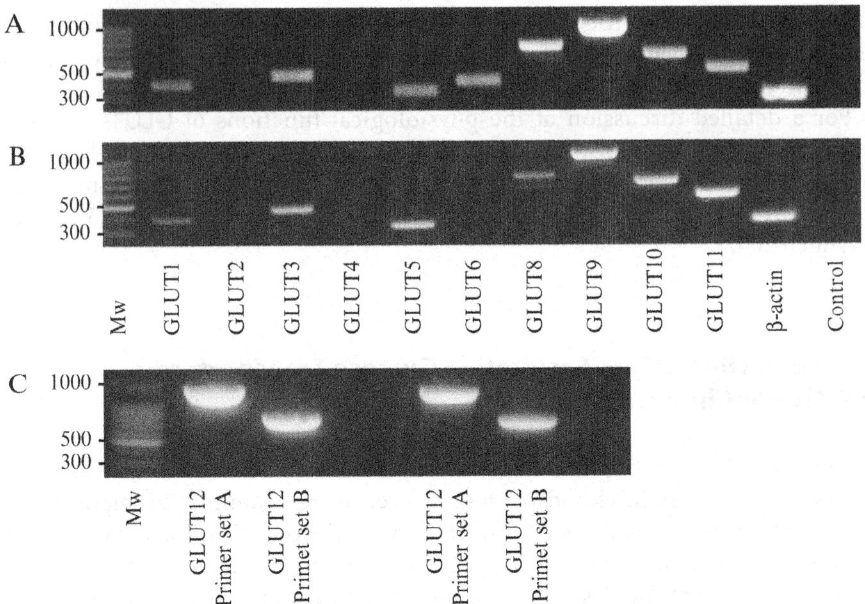

Fig. 8 Evidence for up to nine facilitative glucose transporters in human articular cartilage. **A, B** Summary of the results of RT-PCR experiments using two different human cartilage cDNA libraries which confirm the expression of GLUT1, GLUT3, GLUT5, GLUT6, GLUT8, GLUT9, GLUT10, GLUT11 and GLUT12 in cartilage. The template was omitted in the control lanes shown in **A** and **B**. **C** The presence of GLUT12 was confirmed using two different primer sets. Additional information about the sequences of the isoform specific PCR primers and the sizes of the PCR products may be found in a recent paper (Richardson et al. 2003)

7.1
Functional Significance of GLUT1 and GLUT3 in Articular Chondrocytes: The Developmental Perspective

Recent studies by our group and others suggest that GLUT1 and GLUT3 are present in chondrocytes derived from fully developed human (Fig. 10A, B), porcine (Fig. 10C), equine (Fig. 9), and ovine articular cartilage (Mobasheri et al. 2002b, 2002c; Phillips et al. 2005a; Shikhman et al. 2001a) (Figs. 11 and 12). However, the functional significance of GLUT1 and GLUT3 expression in chondrocytes has not been explored using a comparative physiology approach.

The optimal growth, development, and maintenance of musculoskeletal structures are important for skeletal stability. The availability of glucose and the expression of GLUT proteins in musculoskeletal cells are likely to influence the development of the musculoskeletal structures of load-bearing synovial joints (i.e., articular cartilage, synovium, tendon, and ligament) in all vertebrates. The facilitated transport of glu-

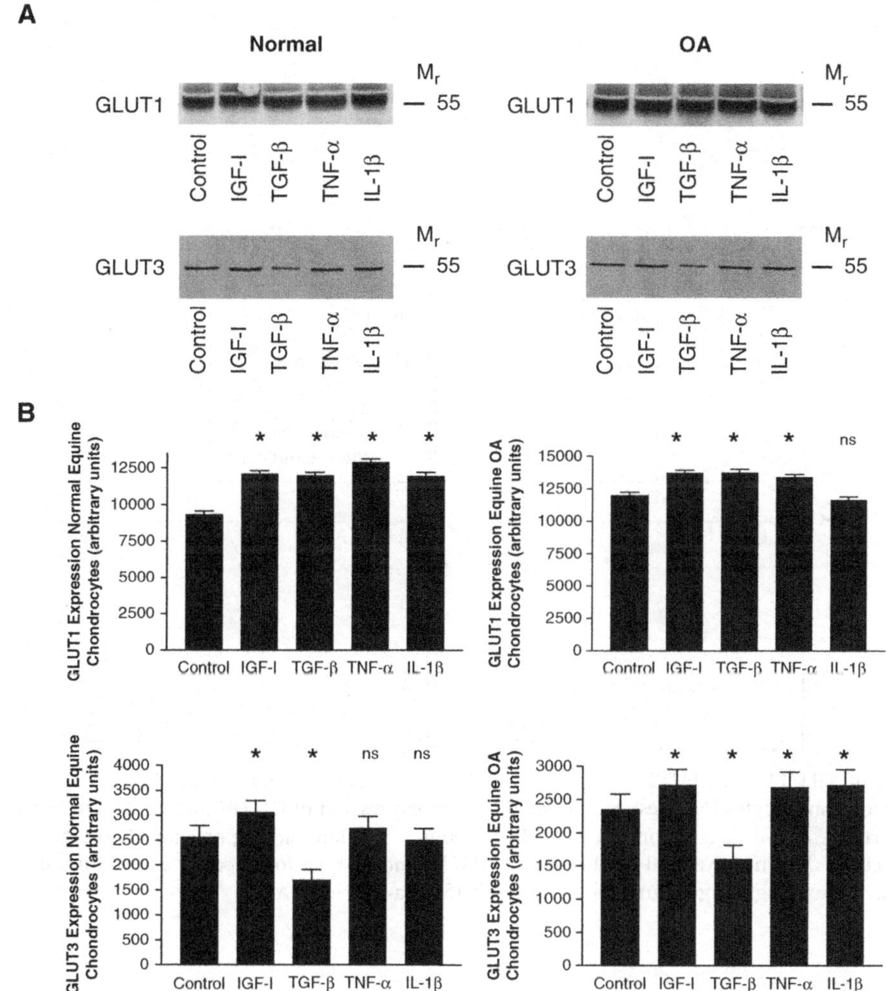

Fig. 9 Evidence for growth factor and cytokine regulation of GLUT1 and GLUT3 expression in normal and osteoarthritic equine articular chondrocytes. Western blot analysis demonstrated that equine chondrocytes express GLUT1 and GLUT3. IGF-I stimulation resulted in upregulation of GLUT1 and GLUT3. TGF-β stimulation upregulated GLUT1 but downregulated GLUT3. TNF-α and IL-1β stimulation resulted in upregulation of GLUT1 but did not affect GLUT3. TGF-β also reduced the levels of the GLUT3 protein. *Denotes a significant difference (P <0.05); *ns* denotes a statistically insignificant difference

cose across the chondrocyte membrane represents the rate-limiting step in glucose metabolism and is essential for the functional integrity of articulating joints (Shikhman et al. 2001a). Thorens and co-workers have reviewed and discussed the role of the mammalian GLUT/SLC2A family of glucose transporters in glucose transport (Joost

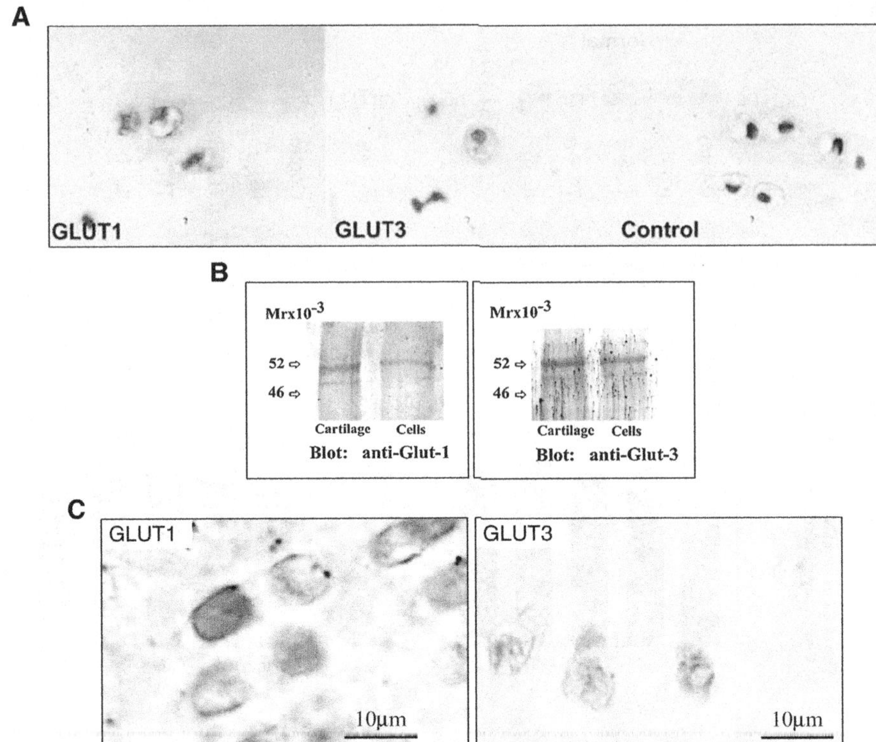

Fig. 10 GLUT1 and GLUT3 are expressed in human articular chondrocytes (**A, B**) and in porcine chondrocytes (**C**). Western blot evidence for expression of GLUT1 and GLUT3 in human
cartilage and isolated chondrocytes (**B**). Sections of human and pig articular cartilage were
incubated with polyclonal antibodies to GLUT1 and GLUT3 followed by an alkaline phos-
phatase-conjugated goat anti-rabbit antibody (Sigma-Aldrich) (**A, C**)

and Thorens 2001; Uldry and Thorens 2004). It is also important to highlight the
fact that the tissue distribution of glucose transporters is not constant throughout
development (Santalucia et al. 1992). High levels of GLUT1 and GLUT3 are present
in a wide range of fetal tissues, but expression of these transporters greatly decreases
after birth in many of these cell types. Abundant levels of the GLUT1 and GLUT3 proteins are present in pre-implantation mouse embryos, since glucose is the main substrate consumed (Pantaleon and Kaye 1998; Pantaleon et al. 2001). During the early
period of organ formation (i.e., brain, heart, skeletal muscle, and kidney) GLUT1 is
primarily responsible for glucose supply to the dividing and differentiating cells (Matsumoto et al. 1995; Santalucia et al. 1992). In the early organogenesis period, high-
affinity glucose transporters may be required because embryonic mammalian cells
may be exposed to hypoxia and may have to rely on anaerobic glycolysis (Matsumoto

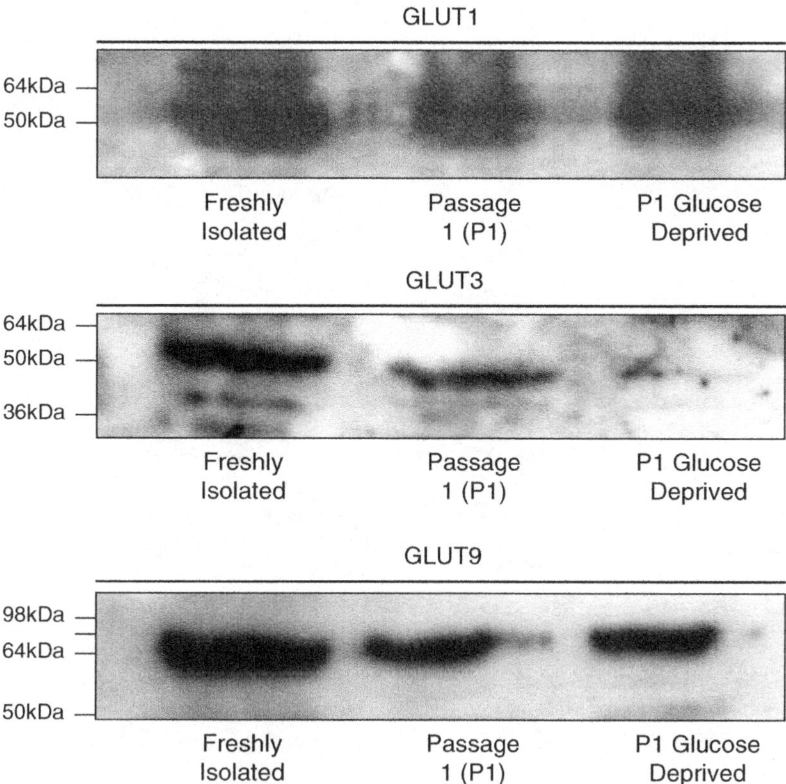

Fig. 11 Expression of GLUT1, GLUT3 and GLUT9 in whole cell lysates of freshly isolated, passage one and glucose deprived ovine articular chondrocytes. Glucose deprivation does not affect expression of GLUT1 but downregulates GLUT3 and GLUT9

et al. 1995). Glucose is particularly important for anabolic activities of the mesenchymal cells that differentiate into specialized cells of the musculoskeletal system (Vannucci and Vannucci 2000). Provision of glucose to growing musculoskeletal tissues is particularly important during fetal development, when cells are rapidly dividing and differentiating (Matsumoto et al. 1995; Pantaleon and Kaye 1998; Santalucia et al. 1992; Vannucci 1994) and involves the GLUT1 isoform (Mobasheri et al. 2005a). Many studies have shown that expression of GLUT1 and GLUT3 is greatly decreased after birth in many cell types. Studies in our laboratory have shown that GLUT1 and GLUT3 are persistently and reproducibly expressed at the mRNA and protein levels in mature chondrocytes derived from the articular cartilage of a variety of species (Mobasheri et al. 2002b, 2002c).

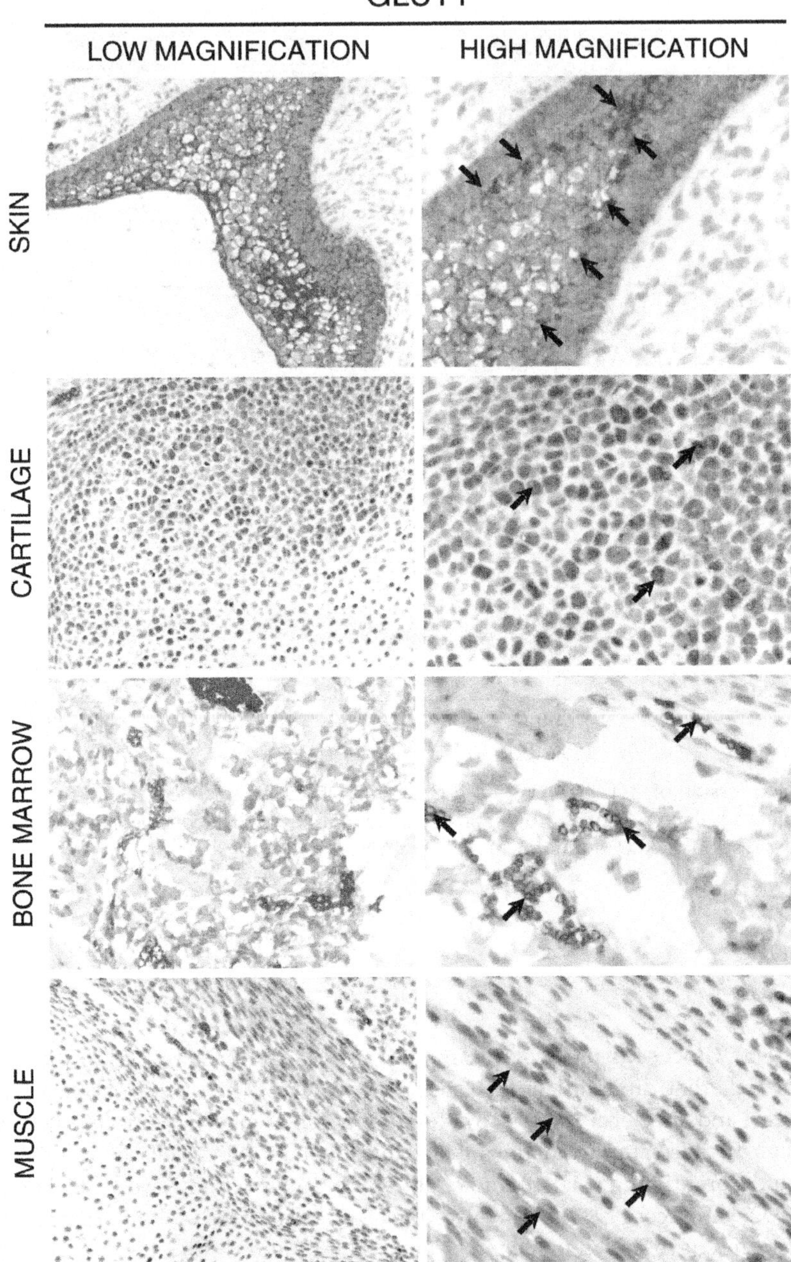

Fig. 12 Expression of GLUT1 in embryonic ovine tissues including skin, cartilage, bone marrow, and skeletal muscle. Data shown are from E42-E45 ovine embryos. See Fig. 33 for more details

7.2
Functional Significance of GLUT1 and GLUT3 in Articular Chondrocytes: The Metabolic Perspective

Studies of glucose transport and metabolism in hypoxia–ischemia in the rat brain have revealed new and valuable information about the specific roles of the GLUT1 and GLUT3 isoforms in regulating glucose uptake in low oxygen conditions. Hypoxia increases the expression of GLUT1 and GLUT3 proteins via an oxygen-sensitive transcription factor (hypoxia-inducible factor 1, HIF-1) to increase glucose transport and the glycolytic rate (Behrooz and Ismail-Beigi 1997, 1999). These proteins could also function as glucose-sensing receptors in tissues exposed to hypoxic conditions.

Cells in the central nervous system are dependent on glucose and oxygen for energy. Neurons in particular need to be buffered from fluctuations in blood glucose. Cerebral hypoxia–ischemia is known to produce major alterations in energy metabolism and glucose utilization in the brain. A number of studies have investigated the effects of hypoxia, glucose deprivation, and hypoxia plus glucose deprivation on the transcription and translation of glucose transporters in neurons (Choeiri et al. 2002) and astroglia (Morgello et al. 1995; Yu et al. 1995). The GLUT1 isoform mediates the transport of glucose across the blood–brain barrier, whereas both GLUT1 and GLUT3 mediate glucose uptake into neurons and glia (Vannucci et al. 1996). Hypoxia–ischemia in the rat brain stimulates upregulation of GLUT1 and GLUT3 glucose transporter gene expression (Vannucci et al. 1998). Animals (or cell lines) treated with the hypoxia mimetic cobalt chloride (a chemical agent that stimulates the expression of a set of hypoxia-responsive genes) also upregulate GLUT1 and GLUT3 expression (Badr et al. 1999).

Other related studies have shown that glucose deprivation alone produces minimal effects on GLUT mRNA levels in the brain, but hypoxia and glucose deprivation synergize to markedly increase GLUT gene expression (Bruckner et al. 1999). Among the various hypoxia-responsive genes, GLUT1 was the first gene whose rate of transcription was shown to be dually regulated by hypoxia (Zhang et al. 1999). It is now well appreciated that GLUT1 and GLUT3 gene expression is acutely regulated by hypoxia-inducible factor 1 (HIF-1).

Our own studies have consistently shown that the hypoxia-responsive GLUT1 and GLUT3 are functionally expressed in equine and ovine chondrocytes (Mobasheri et al. 2002b, 2002c), (Mobasheri et al. 2005a). We have recently presented a new hypothesis that implicates GLUT1 and GLUT3 glucose transporters and the hypoxia-inducible transcription factor (HIF-1α) in glucose sensing in chondrocytes (Mobasheri et al. 2005b). The expression of GLUT1 and GLUT3 in chondrocytes may also be related to the hypoxic nature of cartilage and the unusual metabolic properties of chondrocytes (reviewed in detail in the following section).

7.3
Regulation of Hypoxia and Hypoxia-Responsive Gene Expression by the Transcription Factor HIF-1α in Chondrocytes

Living cells are exposed to a wide spectrum of oxygenation (Fig. 13). HIF-1 is a heterodimeric basic-helix-loop-helix-PAS domain transcription factor that mediates changes in gene expression in response to changes in oxygen concentration (Wang et al. 1995a, 1995b). HIF-1 is the only known mammalian transcription factor expressed uniquely in response to physiologically relevant levels of hypoxia (Iyer et al. 1998; Semenza and Wang 1992). HIF-1 protein is normally degraded under normoxic conditions. However, despite the fact that HIF-1 mRNA levels remain the same in hypoxic conditions, HIF-1 protein is degraded less. HIF-1 activates mRNAs encoding erythropoietin and the following glycolytic enzymes: aldolase, phosphoglycerate kinase, pyruvate kinase, enolase, lactate dehydrogenase, and phosphofructokinase (Semenza et al. 1996, 1994). Other HIF-1 target genes include those encoding vascular endothelial growth factor (VEGF), and the GLUT1 and GLUT3 glucose transporters (Ouiddir et al. 1999; Semenza 2001; Semenza et al. 1999; Vannucci et al. 1998, 1996) (Fig. 13 and 14). Indeed, GLUT1 and GLUT3 are early targets of HIF-1 in hypoxic conditions. The available literature in this area is extensive with much discussion of the role of HIF-1 and glucose transporters in tumor progression and we refer readers to a series of excellent articles by G.L. Semenza (Semenza 2002a, 2002b, 1998, 1999a, 1999b, 2001).

It is now well established that HIF-1α is expressed in chondrocytes and may mediate responses to mechanical overload (Pufe et al. 2004), oxygen deprivation, and metabolic stress (Rajpurohit et al. 2002) by inducing expression of

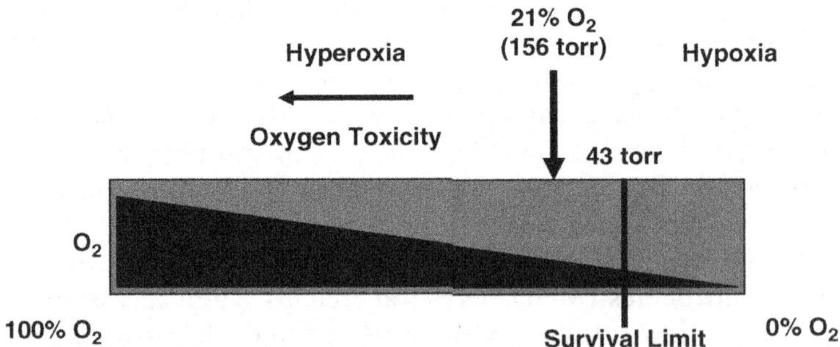

Fig. 13 Physiological and pathophysiological spectrum of oxygenation. Hyperoxia and anoxia are lethal to all cells, whereas most cells have the varied capacity to adapt to low oxygen tension hypoxia. Chondrocytes and mesenchymal stem cells appear to favor hypoxic conditions

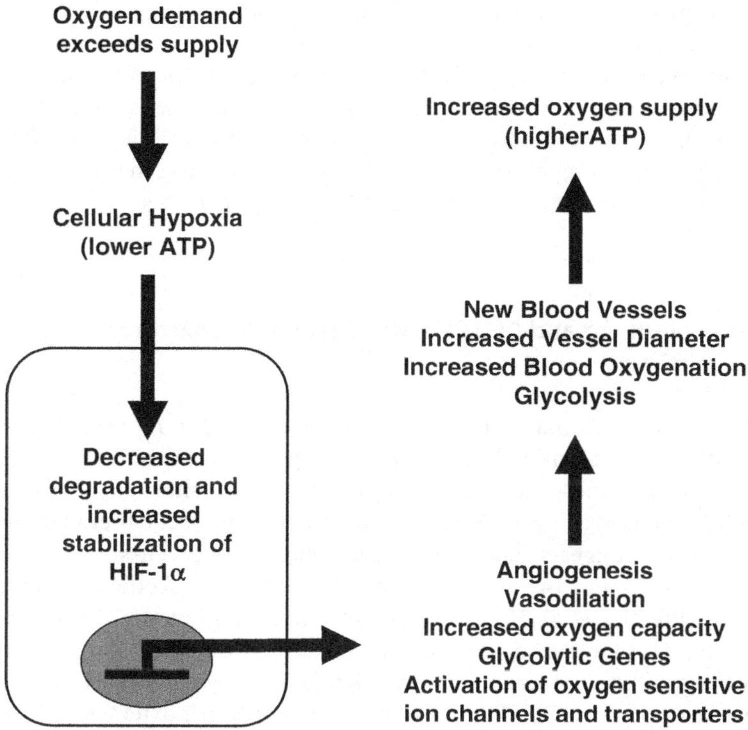

Fig.14 Molecular responses to hypoxia; when oxygen demand exceeds supply the HIF-1α oxygen-sensing system results in activation of key genes involved in angiogenesis, vasodilatation, oxygen delivery, and glycolysis in order to increase tissue oxygenation

VEGF and glucose transporters (Fig. 14). Studies of embryonic and epiphyseal chondrocytes have shown that HIF-1α is essential for chondrocyte growth arrest and survival in vivo (Pfander et al. 2003; Schipani et al. 2001). Recent work on the nucleus pulposus (NP) in the intervertebral disc has also suggested that HIF-1α is important for the maintenance of anaerobic glycolysis and the response to hypoxia and nutrient stress (Pfander et al. 2003; Rajpurohit et al. 2002). HIF-1α is known to upregulate stress-responsive genes and one such gene is the vascular endothelial growth factor (VEGF). Studies of mouse epiphyseal chondrocytes have shown that soluble isoforms of VEGF, VEGF(120) and VEGF(164), are abundantly expressed splice variants in cells exposed to low oxygen levels (Cramer et al. 2004). Thus the biological effects of VEGF in low-oxygen conditions are HIF-1α dependent since functional inactivation of HIF-1α abolishes the hypoxic increase of VEGF expression in chondrocytes (Cramer et al. 2004). HIF-1α may also be involved in the poorly understood process of mechanotransduction; elegant recent studies have shown that mechanical overload increases HIF-1α expression and immunoreactivity in cartilage which, in turn, induces VEGF expression in chondrocytes (Pufe et al. 2004). VEGF is an important growth factor for angiogenesis and vascularization but it also participates in cytokine-mediated inflammatory processes (Pufe et al. 2004, 2001). Therefore, it is becoming clear that HIF-1α-regulated target genes are expressed in chondrocytes and are involved in diverse stress response processes.

7.4
Expression of HIF-1α and GLUT1 in Normal and Osteoarthritic Articular Cartilage

Cells may be exposed to a variety of oxygen tensions ranging from the complete absence of oxygen, or anoxia, to super-atmospheric oxygen concentrations, or hyperoxia; both of these extremes will be toxic in the short term and lethal to living cells over longer periods of time (Fig. 13). The recent literature on cartilage metabolism suggests that chondrocytes and their precursors favor hypoxic conditions (Fig. 13 and 14). Recent immunohistochemical studies by Pfander and co-workers (Pfander et al. 2005) suggest that expression of HIF-1 protein and its target genes GLUT1 and phosphoglycerate kinase 1 (PGK-1) is increased in human chondrocytes with the severity of OA. Work from our own laboratory suggests that the hypoxia-responsive GLUT1 and GLUT3 glucose transporters and the recently described GLUT9 glucose transporter are expressed in human articular chondrocytes (Mobasheri et al. 2005b; Phillips et al. 2005a; Richardson et al., 2003) Fig. 8 and Fig. 15). Our work in human intervertebral disc cells has shown that GLUT1, GLUT3, and GLUT9 are also present in these cells (see subsequent sections). Thus, chondrocytes may depend on the adaptive functions of HIF-1α in degenerate cartilage in order to maintain production of ATP and matrix macromolecules during the course of OA progression.

Our recent work has suggested that chondrocyte adaptation to hypoxia may occur by metabolic alterations including enhancement of the glucose transporting capacity of the cells, increased glycolysis, and lactate production. We have shown that hypoxia and glucose deprivation increase the production of lactic acid and production of the active form of MMP-2 (Mobasheri et al. 2006) (Fig. 16). Upregulation of MMP-2 and the build-up of lactate will have detrimental effects on the ECM. We have proposed that chronic hypoxia may occur in degenerate osteoarthritic joints and the consequent metabolic alterations may contribute to the pathogenesis and progression of OA.

We have observed that the glucose transport is upregulated in cultured equine chondrocytes in response to hypoxia and hypoxia mimetic agents such as cobalt chloride (Mobasheri et al. 2006) (Fig. 17). We have made very similar observations regarding the expression of GLUT1 and its regulation by hypoxia and cobalt chloride in human C-28/I2 chondrocyte-like cells (Fig. 18). GLUT1 is also regulated

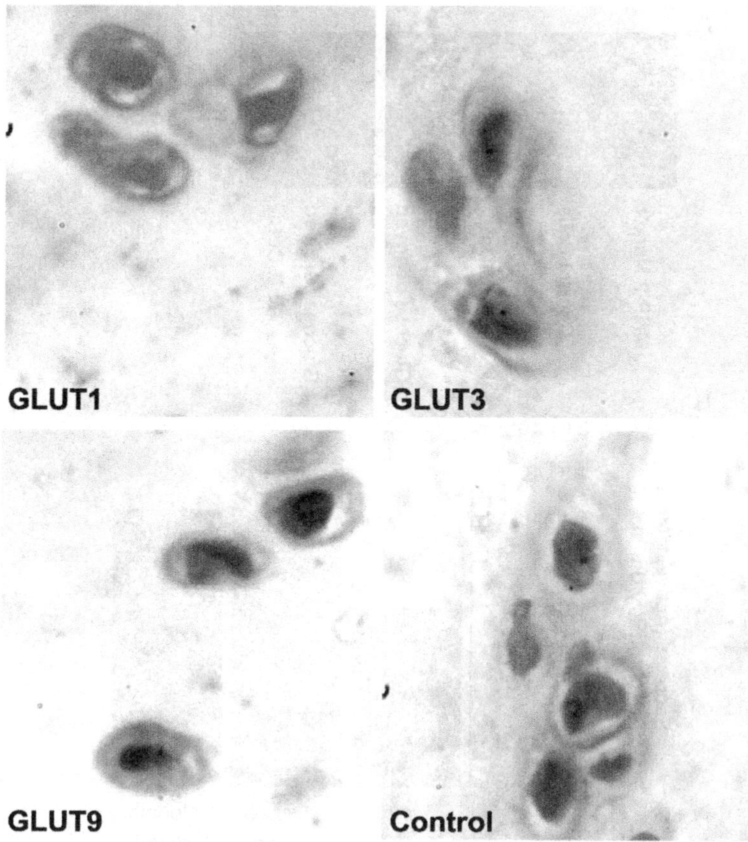

Fig. 15 Immunohistochemical evidence for expression of GLUT1, GLUT3, and GLUT9 glucose transporters in normal human articular cartilage (original magnification ×400). The immunostaining for the GLUT proteins is indicated by the *red* substrate and the cell nuclei were counterstained with hematoxylin. The primary antibody was omitted in the control panel, which shows a section of human cartilage exposed only to secondary alkaline phosphatase conjugated antibody. (Reproduced from Richardson et al. 2003 with copyright permission of Elsevier Science and the OsteoArthritis Research Society International)

by glucose deprivation. As shown in Fig. 18B, GLUT1 levels are higher in C28/Ia chondrocyte-like cells deprived of glucose for 48 h than in cells cultured under normal glucose levels.

We are currently studying glucose uptake in those conditions to determine whether the lower molecular weight GLUT1 form detected is functional.

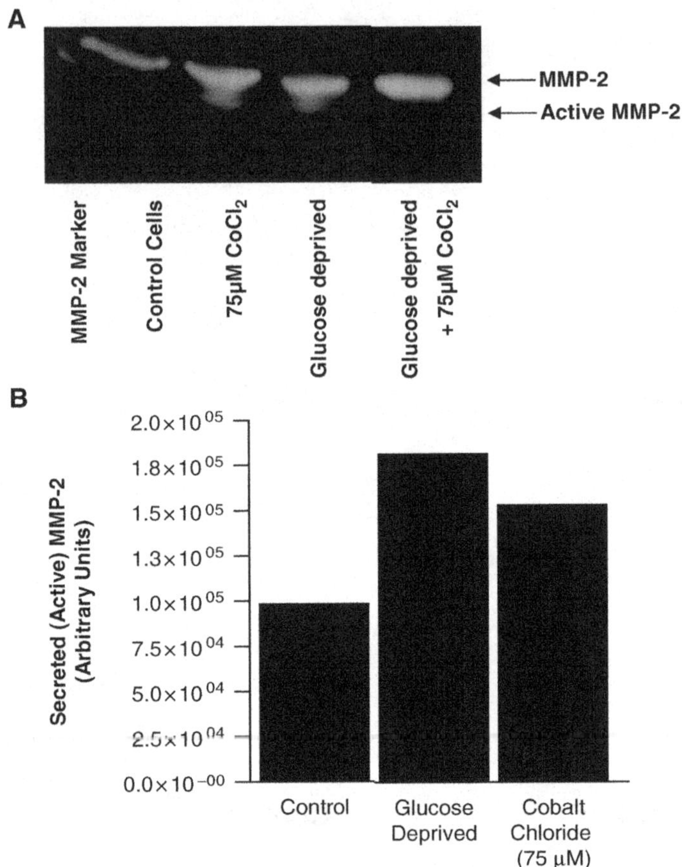

Fig. 16 Effects of glucose deprivation and cobalt chloride on the expression of active MMP-2 secreted into the culture medium of chondrocytes. **A** A representative gelatin zymogram used for the quantitative analysis of active MMP-2 expression. **B** Glucose deprivation, exposure to 75 μM cobalt chloride or a combination of both for periods of up to 24 h significantly increased MMP-2 production secretion by chondrocytes compared to the control group. MMP-2 is detected as two closely migrating bands on the zymogram; the lower molecular weight band corresponds to active MMP-2 and the higher molecular weight band represents inactive MMP-2. (Reproduced from Mobasheri et al. 2006 with copyright permission of the New York Academy of Sciences)

These glucose transporters are also differentially regulated in response to growth factors and proinflammatory cytokines and are probably involved in energy provision for chondrocytes in inflammatory conditions (Mobasheri et al. 2005b, 2002c; Richardson et al. 2003; Shikhman et al. 2001b) (Fig. 19).

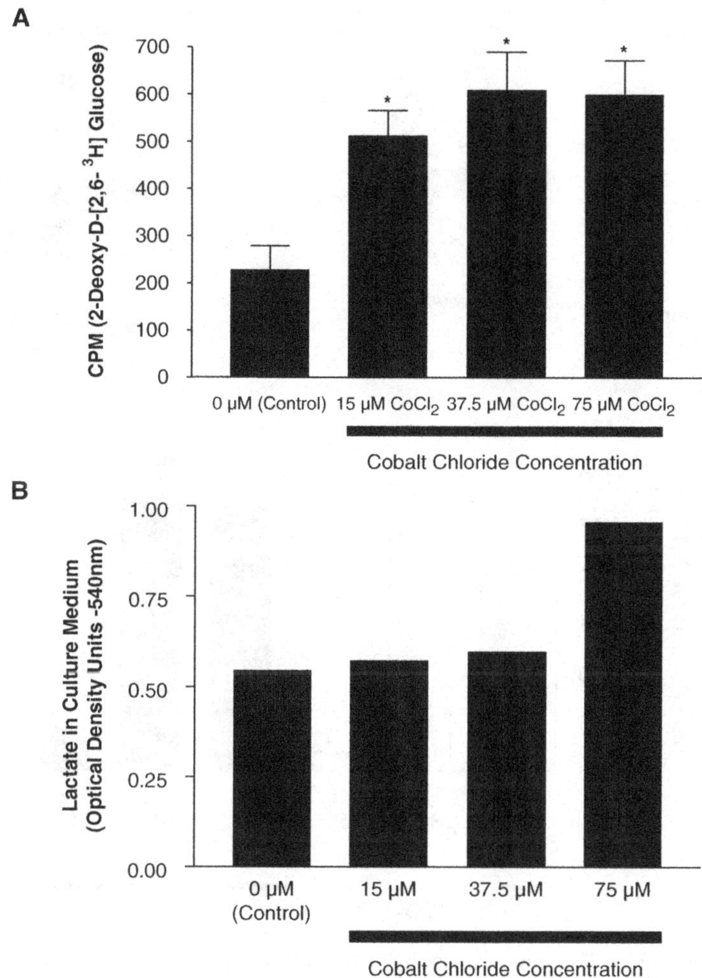

Fig. 17 A Effects of the hypoxia mimetic cobalt chloride on the uptake of 2-deoxy-D-[2, 6-³H] glucose by monolayer cultured equine chondrocytes in 24-well plates. The uptake of 2-deoxy-D-[2, 6-³H] glucose uptake in control chondrocytes was compared with cells incubated for 24 h with increasing concentrations of cobalt chloride (15, 37.5, and 75 μM). The net uptake of 2-deoxy-D-[2, 6-³H] glucose was significantly higher in chondrocytes incubated with cobalt chloride compared to control cells. The highest increase was seen with 75 μM cobalt chloride. *Error bars* indicate standard errors of the means ($n=3$). In cases where a statistically significant difference between the experimental group and a control group was found the bar is labeled with *. **B** Effects of cobalt chloride on the production of lactic acid by monolayer cultured equine chondrocytes in 24-well plates. The production of lactate in culture supernatants of control chondrocytes was compared with cells incubated for 24 h with increasing concentrations of cobalt chloride (15, 37.5, and 75 μM). Lactic acid production was higher in chondrocytes incubated with cobalt chloride compared to control cells and the effect seemed to plateau at a concentration of 75 μM cobalt chloride. (Reproduced from Mobasheri et al. 2006 with copyright permission of the New York Academy of Sciences)

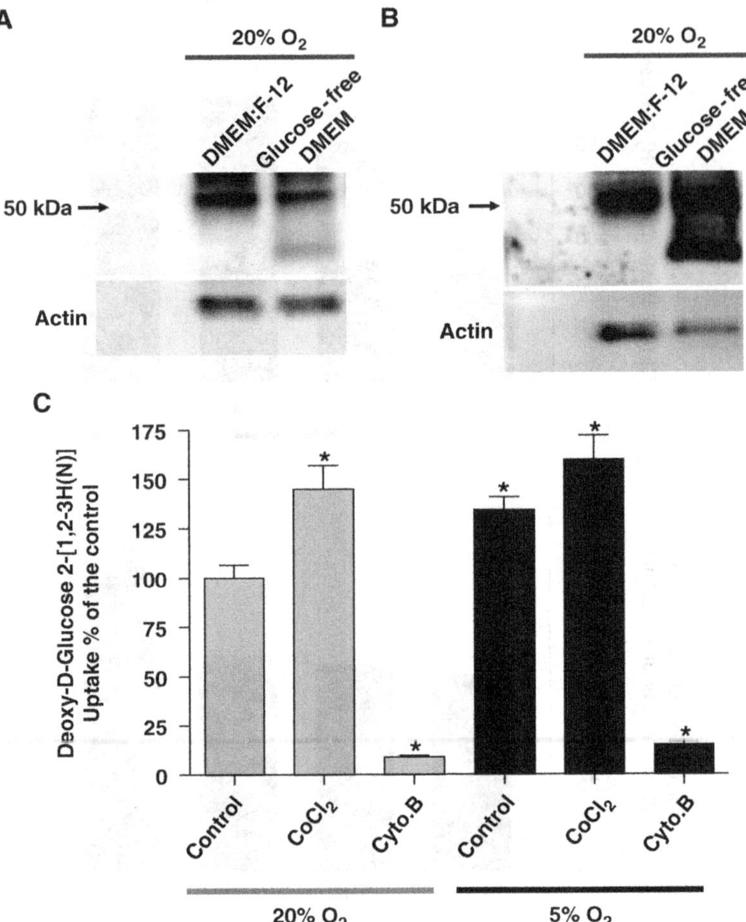

Fig. 18 Expression of GLUT1 and its regulation by glucose deprivation, hypoxia and cobalt chloride (CoCl$_2$) in human C-28/I$_2$ chondrocyte-like cells. The GLUT1 protein was detected as an approximately 50 kDa protein in whole cell extracts of C-28/I2 chondrocyte-like cells. In cells incubated for 24 (**A**) or 48 h (**B**) in the absence of glucose, an additional band was detected corresponding to approximately 40 kDa. **C** Net glucose transport was significantly higher in chondrocytes maintained in hypoxia (135.33% increase compared to normoxia controls, ±3.5, n =4); exposure of C-28/I$_2$ chondrocytes to CoCl$_2$ both in normoxia (153.24%, ±16.7, n=4) and hypoxia (159.68% increase compared to the controls, ±14.1, n =4) conditions also resulted in a significant increase in net glucose uptake. Statistical analysis was carried out using the one-way ANOVA test and the means were found to be significantly different. (P =0.0107, significance level, P < 0.05)

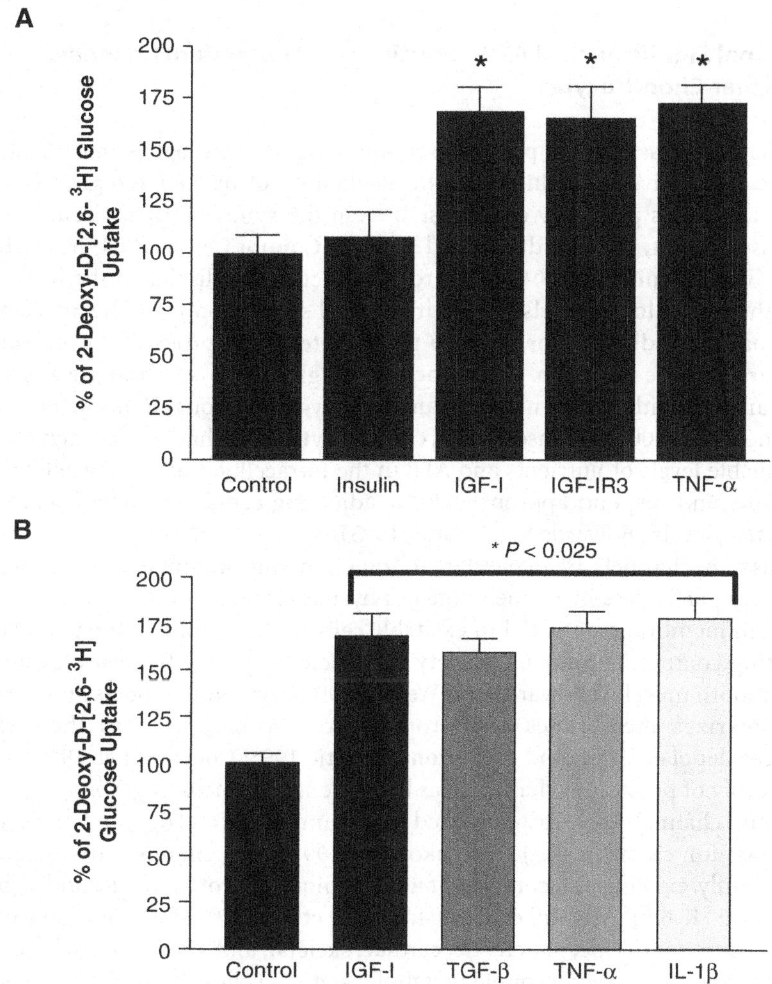

Fig. 19 A Effects of insulin, insulin-like growth factor (IGF-I) and TNF-α on 2-deoxy-D-[2, 6-³H] glucose uptake by C20/A4 chondrocytes. Chondrocytes were stimulated with insulin (12.5 μg ml⁻¹), TNF-α (20 ng ml⁻¹), and recombinant (long R3) IGF-I (20 ng ml-1) for a period 24 h at 37°C before facilitated glucose transport was measured by 2-deoxy-D-[2, 6-³H] uptake. Baseline 2-deoxy-D-[2, 6-³H] uptake in un-stimulated chondrocytes (control) was considered as 100%. *Error bars* indicate standard errors of the means ($n = 3$). In cases where a statistically significant difference between the experimental group and a control group was found the bar is labeled with *. (Reproduced from Richardson et al. 2003 with copyright permission of Elsevier Science and the OsteoArthritis Research Society International). **B** Uptake of 2-deoxy-[2, 6-³H] glucose by equine articular chondrocytes stimulated with IGF-I, TGF-β, TNF-α and IL-1β. Baseline 2-deoxyglucose uptake in un-stimulated chondrocytes (control cells) was considered as 100%. *Error bars* indicate standard errors of the means ($n = 3$). An *asterisk* denotes a significant difference between control and experimental groups ($P < 0.025$). (Reproduced from Phillips et al. 2005b with copyright permission of Elsevier Science)

7.5
Functional Significance of ATP-Sensitive (K_{ATP}) Potassium Channels in Articular Chondrocytes

As we have discussed in the previous section, articular cartilage is an avascular and hypoxic connective tissue in which the availability of oxygen and glucose is limited and depends primarily on diffusion from the synovial microcirculation and, to a lesser extent, subchondral blood vessels (Coimbra et al. 2004; Mobasheri et al. 2005b; Schipani et al. 2001). In a previous section we discussed the importance of subchondral blood vessels and their physical separation from the articular cartilage making it difficult for them to participate in the provision of nutrients to articular cartilage nutrition. Chondrocytes are glycolytic cells and are able to survive in an ECM with limited nutrients and low oxygen tensions (Henrotin et al. 2005a; Mobasheri et al. 2005b). Consequently, chondrocytes must have the capacity to sense the available levels of nutrients and ATP in the intracellular and extracellular compartments and respond appropriately by adjusting cellular metabolism and ATP production levels (Edwards and Weston 1995).

Potassium channels are integral membrane proteins present in most mammalian cells. They participate in a wide range of physiological responses including control of the cell membrane potential in excitable cells in the brain and the pancreas and regulating contractile tone in a variety of muscle types (cardiac, skeletal, and vascular smooth muscle) (Edwards and Weston 1995). Opening of potassium channels hyperpolarizes membranes and promotes quiescence, whereas their closure produces depolarization and excitation (Christie 1995; Coetzee et al. 1999). A large superfamily of potassium channels has been identified, including: voltage-activated potassium channels (K_v), Ca^{2+}-activated potassium channels (K_{Ca}), and inward rectifier potassium channels (K_{ir}) (Babenko et al. 1998). K_{ATP} channels are members of the K_{ir} family, existing as a complex of an ATP-binding protein (SUR) and a channel component ($K_{ir}6.x$) (Minami et al. 2004; Quayle et al. 1997). They are widely distributed in neuroendocrine, pancreatic, cardiac, skeletal, and smooth muscle cells; it is believed that they serve to couple metabolic state to cellular activity. This coupling is important in both physiological and pathological conditions (Dart and Standen 1993, 1995).

Recent studies on ion channels in chondrocytes have expanded our understanding of the roles of these proteins in chondrocyte and cartilage function. Nevertheless, data on chondrocyte ion channels are still very limited compared to other well-studied cell types. Despite the physiological importance of potassium channels in the modulation of metabolic activity, there are significantly fewer publications relating to potassium channel expression and function in chondrocytes compared to other tissues. Low oxygen tension and hypoxia are known to lead to activation of K_{ATP} channels in other systems (Phillis 2004; Dart and Standen 1993, 1995) suggesting that these channels are important in hypoxia-mediated cell signaling (Mobasheri et al. 2007). In a recent study we used a combined electrophysiological and immunohistochemical approach to test the hypothesis that K_{ATP} channels are expressed in articular chondrocytes and may therefore be involved in metabolic regulation. We used the patch-clamp technique to investigate whether K_{ATP} channels are functionally expressed in isolated equine articular chondrocytes and employed immunohistochemistry to determine the expression of $Kir_{6.1}$ in human and equine chondrocytes from both normal and OA cartilage (Mobasheri et al. 2007). The results of this study show, for the first time, that K_{ATP} channels ($Kir_{6.1}$ subunit of the channel) are present in normal and OA chondrocytes from equine and human subjects (Fig. 20). Furthermore, we have demonstrated that K_{ATP} channels are functionally expressed in equine chondrocytes (Fig. 21). In view of their function in other cell types (Fig. 22), we have proposed that these potassium ion channels may be important in the regulation of cartilage metabolism and intracellular ATP sensing (Fig. 23) (Mobasheri et al. 2007; Pfander et al. 2003; Rajpurohit et al. 2002; Schipani et al. 2001). In the pancreatic β cell the K_{ATP} channel senses metabolic changes in the cell, thereby coupling metabolism to electrical activity and ultimately to insulin secretion (Ashcroft and Gribble 1998). When K_{ATP} channels open, β cells hyperpolarize and insulin secretion is suppressed. We propose that in a situation

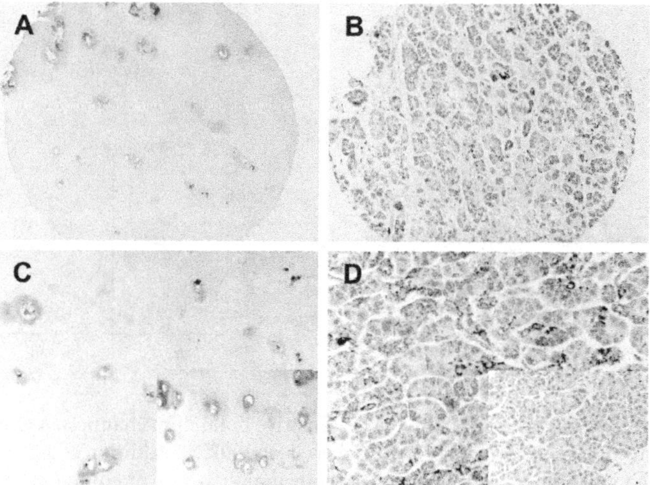

Fig. 20 $Kir_{6.1}$ is expressed in chondrocytes in normal human articular cartilage. Immunohistochemical analysis of $Kir_{6.1}$ expression in samples of full-depth human articular cartilage and human pancreas represented on the CHTN2002N1 multiple human Tissue MicroArrays (TMAs). Incubation of CHTN2002N1 TMAs with polyclonal antibodies to $Kir_{6.1}$ followed by horseradish peroxidase-labeled rabbit anti-goat IgG (DakoCytomation) produced positive immunostaining in chondrocytes in human knee cartilage (low magnification shown in **A**, high magnification shown in **C**) and human pancreas (low magnification shown in **B**, high magnification shown in **D**). Omission of primary antibody from the immunohistochemical procedure resulted in complete abrogation of specific immunostaining of human articular chondrocytes in cartilage samples (*inset*, **C**) and pancreatic cells (*inset*, **D**). (Reproduced from Mobasheri et al. 2007 with copyright permission of Elsevier Science and the OsteoArthritis Research Society International)

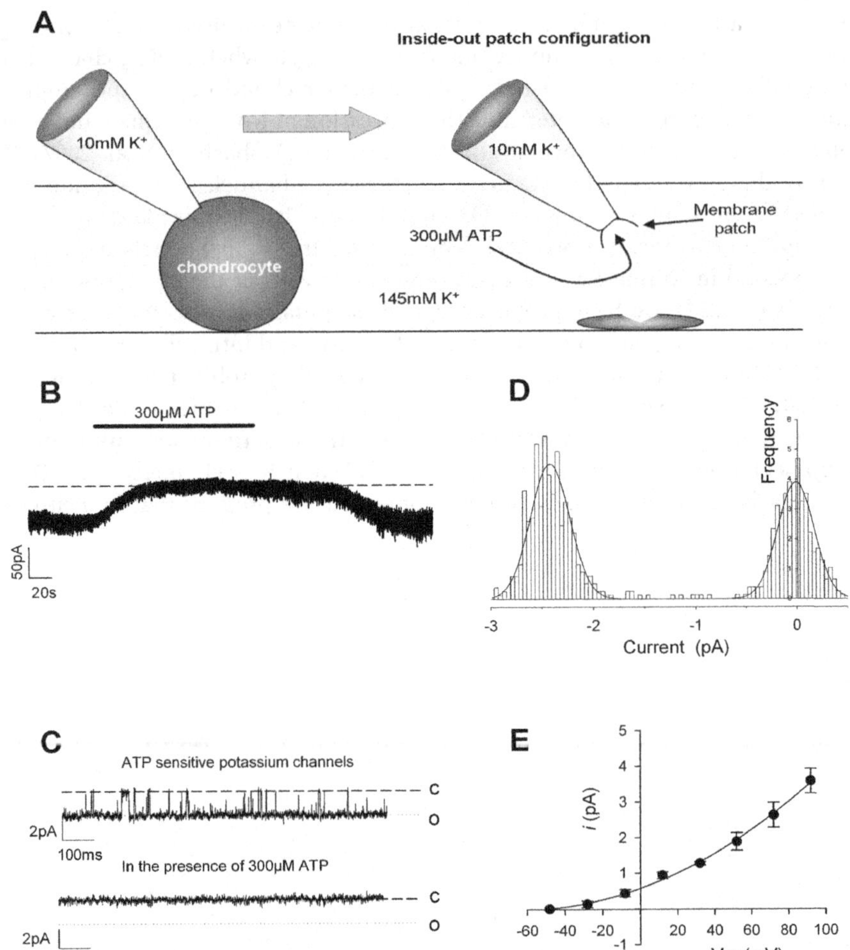

Fig. 21 Chondrocytes express functional ATP-sensitive potassium channels. **A** Outline of experimental design, for further details see (Mobasheri et al. 2007). **B** Addition of 300 µM ATP (added by bath perfusion) inhibits currents in inside-out maxi-patches of equine articular chondrocytes. *Dashed line* indicates the zero current level, holding potential −60 mV. **C** Inside-out single-channel patch in the absence (*upper panel*) and presence (*lower panel*) of 300 µM ATP. The *dashed line* represents the zero current level, where the channel is closed ("*C*") and the *dotted line* represents the unitary current level, where the channel is open ("*O*"). Holding potential −60 mV. **D** All points amplitude histogram from the patch shown in **C**. **E** Current–voltage curve for ATP-sensitive potassium channels recorded in a number of experiments similar to **C** and **D**. (Reproduced from Mobasheri et al. 2007 with copyright permission of Elsevier Science and the OsteoArthritis Research Society International)

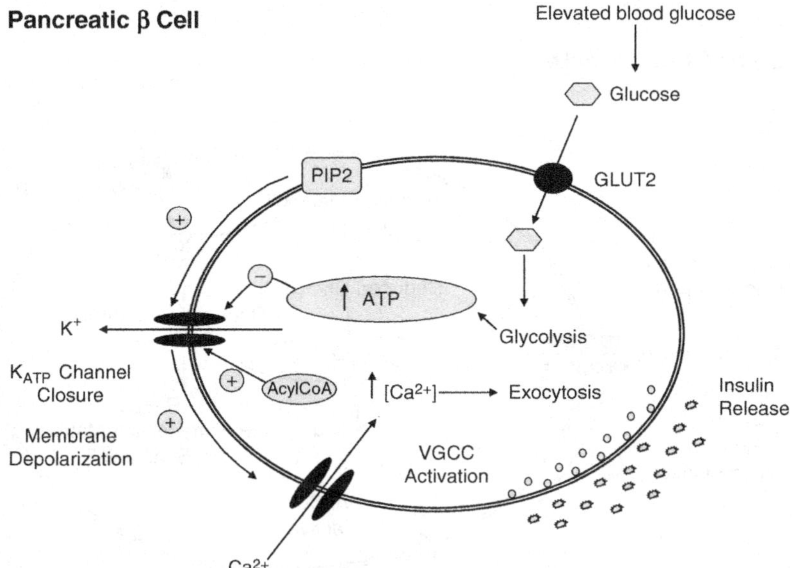

Fig. 22 The role of the pancreatic β cell K_{ATP} channel in secretion of insulin in response to elevated blood glucose levels. A rise in blood glucose is an important metabolic signal that closes K_{ATP} channels, causing membrane depolarization, activation of voltage gated calcium channels (VGCC), free calcium entry and insulin release by exocytosis. It is thought that various additional effectors including phosphatidylinositol-4,5-bisphosphate (PIP_2) and acyl CoAs modulate the ATP sensitivity of the K_{ATP} channel thereby affecting the coupling of pancreatic cell metabolism to insulin secretion. (Adapted from Koster et al. 2005)

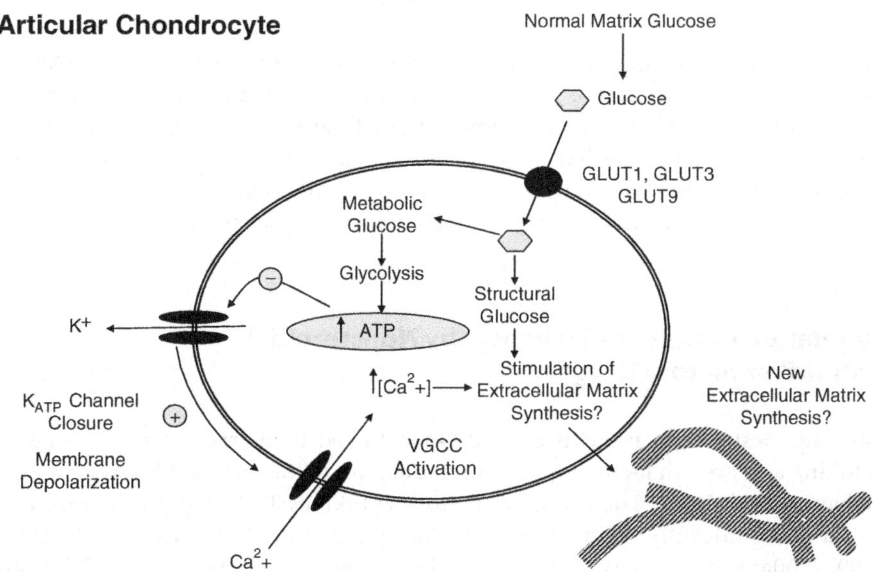

Fig. 23 Proposed role of the chondrocyte K_{ATP} channel in metabolic regulation and intracellular ATP sensing. In this hypothetical model the K_{ATP} channel may be part of extracellular glucose and intracellular ATP sensing machinery of the chondrocyte. Adequate provision of glucose ensures optimal glucose levels for the distinct metabolic and structural pools of glucose within chondrocytes which will promote anabolic processes including extracellular matrix synthesis

Articular Chondrocyte

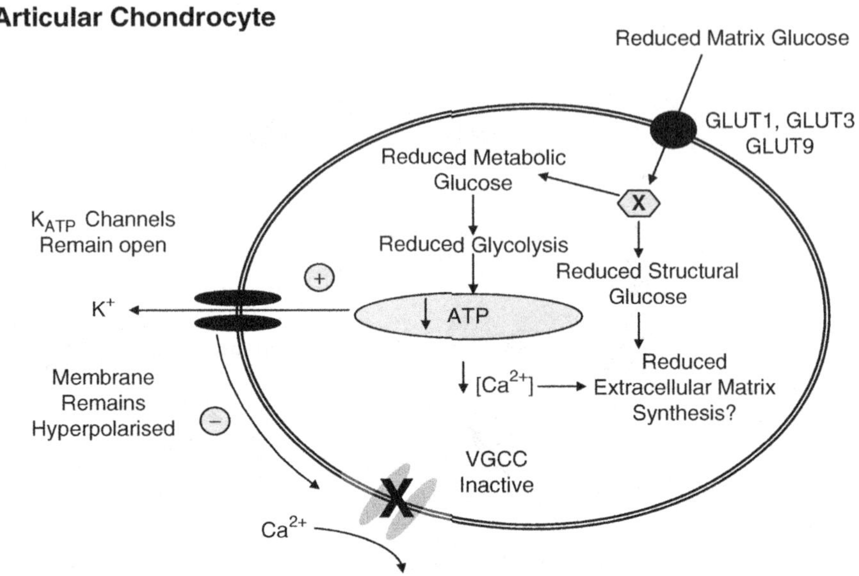

Fig. 24 Putative contribution of the chondrocyte K_{ATP} channel to sensing reduced matrix and intracellular ATP levels. In this scenario a deficiency or shortfall in glucose delivery to chondrocytes may impede anabolic processes and result in a net reduction in extracellular matrix synthesis

where the supply of glucose to the synovial joint is affected (i.e., in metabolic disease), this may affect the process of metabolic regulation which may reduce ECM synthesis or affect matrix turnover (Fig. 24). Further experiments are underway to investigate the mechanisms of extracellular glucose and intracellular ATP sensing by chondrocytes and whether these are perturbed in disease, thus impairing optimal metabolic regulation and extracellular glucose sensing.

8
Regulation of Glucose Transport by Nonsteroidal Anti-inflammatory Drugs

Cartilage destruction in arthritis and OA is linked to aberrant proinflammatory cytokine and growth factor expression in the joint (Chikanza and Fernandes 2000; Malemud et al. 2003). The proinflammatory cytokines TNF-α and IL-β have been found in significantly elevated levels in the synovial fluid of OA joints (Goldring 1999, 2000a; van den Berg 1999). Catabolic pathways are activated by TNF-α and IL-β, which are both upregulated in OA (Malemud et al. 2003). These proinflammatory mediators cause an increase in cartilage matrix degradation through increased MMP and aggrecanase activity in the joint. In addition, TNF-α and

IL-β downregulate ECM protein biosynthesis while concomitantly up-regulating matrix MMP gene and protein expression. When MMPs are activated, cartilage ECM degradation ensues apparently because levels of endogenous cartilage MMP inhibitors cannot regulate MMP activity (Malemud et al. 2003).

TNF-α and IL-β are also important regulators of facilitated glucose transport and metabolism in chondrocytes and synovial fibroblasts and may be implicated in the latter stages of degenerative disorders of articulating joints such as OA. Studies on chondrocytes have shown that IL-β increases glucose uptake by chondrocytes and this is inhibited by cortisol (Hernvann et al. 1996). Higher basal 2-deoxy-D-glucose uptake has also been observed in rheumatoid synovial cells compared to nonrheumatoid synovial cells and this was found to be associated with an increased IL-1β secretion (Hernvann et al. 1992). Work from our laboratory and other groups has shown that TNF-α and IL-β increase net glucose transport, a process that is now generally accepted to be mediated by members of the GLUT/SLC2A family of sugar/polyol transport facilitators expressed in articular chondrocytes and synoviocytes (Hernvann et al. 1992, 1996; Mobasheri et al. 2002c; Phillips et al. 2005b; Shikhman et al. 2001a). The increase in glucose uptake in response to cytokine stimulation appears to be the result of upregulation of glucose transporters in these cells, a process mediated by PKC and p38 MAP kinase activation (Shikhman et al. 2004).

We have recently established and used an in vitro model of primary equine chondrocytes in monolayer to study the effects of proinflammatory cytokines and nonsteroidal anti-inflammatory drugs (NSAIDs) on chondrocytes glucose transport. In particular we were interested in comparing the effects of human recombinant proinflammatory cytokines on glucose transport in equine chondrocytes and the effects of NSAIDs on glucose transport and metabolism. We were also interested in comparing glucose uptake in chondrocytes isolated from normal and OA joints. Therefore, we used this model to compare 2-deoxy-D-glucose uptake in freshly isolated equine chondrocytes from normal and OA equine joints. We hypothesized that chondrocytes isolated from equine OA joints would exhibit increased glucose transport compared to chondrocytes from normal joints. We measured the specific uptake of nonmetabolizable 2-deoxy-D-[2,6-³H] deoxyglucose into normal, cytokine-stimulated (TNF-α; IL-1β; oncostatin M, OSM) and OA chondrocytes from equine joints and assessed the effects of Meloxicam (Mobic®), an NSAID and preferential COX-2 inhibitor, on glucose uptake in normal and cytokine-stimulated cells. Equine chondrocytes were isolated from normal and OA joints by collagenase digestion and grown in monolayer culture with 4% fetal calf serum for no more than two passages. The uptake of [2,6-³H] deoxyglucose into normal and OA chondrocytes from equine joints was assayed as previously described (Phillips et al. 2005b; Richardson et al. 2003). Glucose uptake was also compared with unstimulated C20/A4 human chondrocyte-like cells. Glucose uptake experiments were also performed with equine chondrocytes stimulated for 48 h with human recombinant TNF-α (10 ng/ml), human recombinant IL-1β (10 ng/ml), human recombinant OSM (10 ng/ml), and Meloxicam (5 μg/ml, physiological concentration).

The rate of glucose transport into chondrocytes isolated from equine OA joints was significantly higher than cells derived from normal equine joints and human

C20/A4 chondrocyte-like cells which were used as a reference (Fig. 25A). TNF-α and IL-β significantly increased bulk glucose uptake in equine articular chondrocytes (up to 325% and 318% respectively) but no alteration in glucose transport was seen with human recombinant OSM (Fig. 25B). Treatment with the COX-2 inhibitor Meloxicam resulted in a modest decrease in glucose transport in TNF-α- and IL-β-stimulated chondrocytes, but no significant effect was seen with control and OSM-treated cells (Fig. 25B). The net uptake of 2-deoxy-D-glucose was used as a surrogate marker of chondrocyte glucose transport and metabolic activity. The effects of Meloxicam, a preferential COX-2 inhibitor, on glucose transport were monitored in both stimulated and unstimulated cells. We found that equine

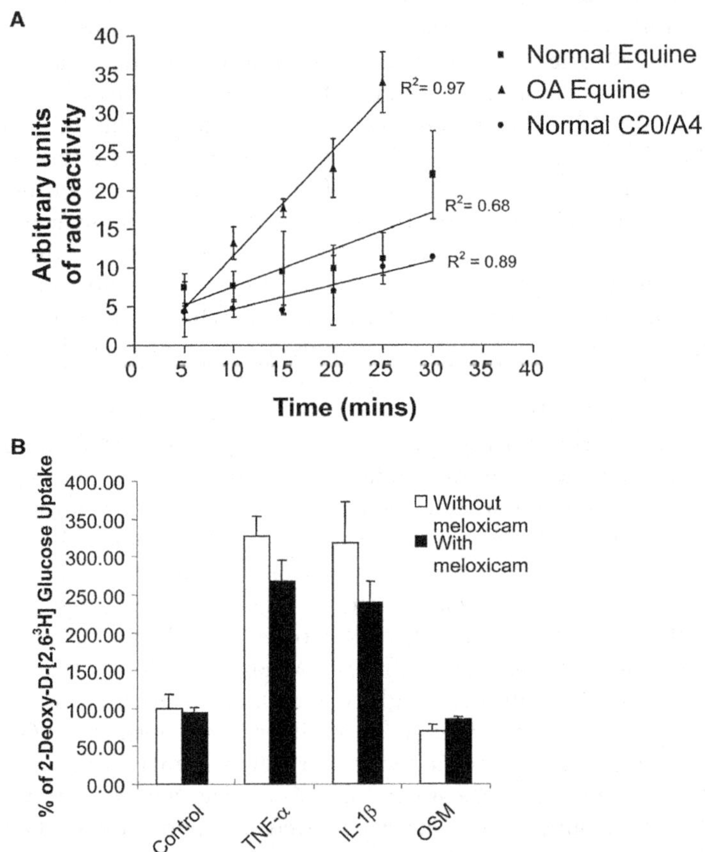

Fig. 25A 2-Deoxy-D-[2, 6-^{3}H] glucose uptake is higher in chondrocytes from inflamed OA joints compared to chondrocytes from normal joints and C20/A4 chondrocyte-like cells. **B** The non-steroidal anti-inflammatory drug Meloxicam reduces 2-deoxy-D-[2, 6-^{3}H] glucose uptake in chondrocytes stimulated with TNF-α and IL-β. Incubation with oncostatin M (OSM) has no effect on 2-deoxy-D-[2, 6-^{3}H] glucose uptake and Meloxicam also has no effect on this uptake

chondrocytes isolated from OA joints exhibited a higher capacity for net glucose uptake compared with equine chondrocytes isolated from normal joints and the C20/A4 human chondrocytes-like cell line.

As expected and previously observed (Hernvann et al. 1992, 1996; Richardson et al. 2003; Shikhman et al. 2001a), both proinflammatory cytokines increased glucose uptake in equine articular chondrocytes. In contrast, no alteration in glucose transport was seen with human recombinant oncostatin M. Incubation with the COX-2 inhibitor Meloxicam appeared to have a modest modulatory effect on glucose uptake (Fig. 25B), reducing glucose uptake in equine chondrocytes treated with TNF-α and IL-1β).

This preliminary study demonstrates that TNF-α- and IL-1β-stimulated chondrocytes and cells derived from equine OA joints transport increased quantities of glucose compared with OSM-treated cells and cells from normal joints. Meloxicam (Mobic®) is prescribed to reduce pain and inflammation in OA and rheumatoid arthritis. The Meloxicam data suggest that this drug appears to have the capacity to reduce glucose transport, which may be a beneficial cellular effect as increased glucose uptake is needed to sustain inflammation in OA. Meloxicam has been reported to influence cartilage metabolism to varying degrees. Blot and co-workers (Blot et al. 2000) showed that Meloxicam increases the synthesis of proteoglycans and hyaluronan in human cartilage with OA in a dose-dependent manner. Studies by other investigators suggest that Meloxicam is a potent inhibitor of collagenase enzyme activity (Barracchini et al. 1998, 1999). However, Meloxicam did not inhibit the in vitro biosynthesis of sulfated proteoglycans in human cartilage tissue from subjects with OA or in normal porcine cartilage (Rainsford et al. 1997). In agreement with this, Rainsford and colleagues have shown there were no effects of Meloxicam on synthesis of proteoglycans and glycosaminoglycans in canine articular cartilage (Rainsford et al. 1999). Our results suggest that Meloxicam exerts modest effects on the uptake and transportation of glucose across the cell membrane and although it is unable to block TNF-α- and IL-1β-mediated stimulation of glucose uptake in chondrocytes, it is able to reduce the uptake of glucose in cytokine-treated chondrocytes. Glucose transporters represent the rate-limiting step for glucose uptake across the chondrocyte membrane (Shikhman et al. 2001a). In cytokine-stimulated chondrocytes they not only maintain but also potentially increase the steady supply of glucose that is required for inflammation. Inhibiting the uptake of glucose in inflammatory situations may therefore represent an unexpected but beneficial side effect of NSAIDs such as Meloxicam, which is one of the most commonly used drugs for the treatment of OA in horses.

9

Glucose Transporters in the Intervertebral Disc

The intervertebral disc is a cartilaginous structure that resembles articular cartilage in its biochemistry and cell biology, but morphologically it is clearly different (Roberts 2002; Urban and Roberts 2003). The disc shows degenerative and ageing changes earlier than any other connective tissue in the human body (Urban and Roberts 2003). The maintenance of the ECM in the NP of the adult human disc is dependent on the functional integrity of the cartilage end plate cells (Pritzker 1977). Cartilage end plate senescence is followed by compensatory cartilaginous metaplasia of annulus fibrosus cells. It has been proposed that intervertebral disc narrowing and collapse are related to metabolic failure of matrix production by end plate and annulus fibrosus cells (Pritzker 1977).

As we have already discussed in earlier sections, degenerative disorders of the intervertebral disc (IVD) and articular cartilage are generally characterized by disequilibrium between ECM repair and degradative processes (Freemont et al. 2002; Le Maitre et al. 2004a, 2004b, 2005). Molecular alterations include elevated MMP (Crean et al. 1997; Goupille et al. 1998; Le Maitre et al. 2004a) and aggrecanase activity (Le Maitre et al. 2007b), increased expression of (and sensitivity to) catabolic cytokines (Le Maitre et al. 2005), and disc cell senescence (Le Maitre et al. 2007a) and apoptosis (Gruber and Hanley 1998). These alterations lead to narrowing of the disc space (Yasuma et al. 1990) and mechanical failure of the disc (Iatridis et al. 1999a, 1999b; Thompson et al. 2000). Cell density in the fully developed, healthy IVD is very low (approximately 4,000/mm^3 in the NP) (Maroudas et al. 1975) and the poor diffusion of nutrients (i.e., O_2 and glucose) and accumulation of metabolic waste products such as lactate (Bartels et al. 1998; Ohshima and Urban 1992) present disc cells with further environmental challenges because of the absence of microvasculature (Repanti et al. 1998). Normally the cells of the outer annulus fibrosus (OAF) are supplied with nutrients by the blood vessels within the outer surface of the AF, but the cells of the inner annulus fibrosus (IAF) and NP may be up to 8 mm away from the nearest capillary and their nutrients must be supplied by diffusion through the disc matrix from blood vessels that terminate in the vertebrae above and below the disc (Bibby et al. 2001; Urban et al. 2001). Studies on the metabolism of the intervertebral disc and articular cartilage suggest that these connective tissues are metabolically and bioenergetically different (Lee and Urban 1997, 2002). In the intervertebral disc, anoxia and the use of glycolysis inhibitors result in a progressive 'positive Pasteur effect' suggesting that a large proportion of the intervertebral disc's energy is derived from oxidative phosphorylation (Ishihara and Urban 1999). This situation is the opposite of that seen in articular cartilage in which anoxia severely inhibits glucose uptake and lactate production and the decrease in lactate formation correlates well with decreased glucose uptake by chondrocytes. This reduction in the rate of glycolysis in anoxic conditions is seen as evidence for a 'negative Pasteur effect' in cartilage (Lee and Urban 1997). These studies have led to suggestions that cell metabolism is compromised by nutrient and oxygen deprivation affecting energy transduction in the disc.

Recent studies on the basic cell biology of rat disc cells have constructed a 'metabolic' and 'phenotypic signature' for cells of the NP based on the hypothesis that in response to oxygen and nutrient deprivation, NP cells express three key phenotypic markers—HIF-1α, GLUT1, and MMP2 (Rajpurohit et al. 2002)—which enables them to be distinguished from surrounding tissues such as the IAF and OAF.

HIF-1α has also been recently demonstrated in the NP of normal human IVD (Ha et al. 2006). The presence of these proteins in the NP has hinted that cells in avascular and hypoxic regions of the disc are capable of adaptive responses to low oxygen tensions and nutrient deprivation. An integral part of the metabolic adaptation to hypoxia is activation of genes involved in promoting anaerobic glycolysis (i.e., hypoxia-responsive GLUT1 and GLUT3 facilitative glucose transporters) (Vannucci et al. 1998), lactate dehydrogenase (LDH), and phosphofructokinase (PFK) (Mobasheri et al. 2005b; Richardson et al. 2003; Semenza et al. 1996, 1994).

Based on the above observations and our recent studies of glucose transport in articular chondrocytes (Mobasheri et al. 2002b, 2002c; Richardson et al. 2003), we hypothesized that the hypoxia-responsive glucose transporters GLUT1 and GLUT3 are expressed in intervertebral disc cells and their expression is increased in disc degeneration via activation of the HIF-1α transcription factor. We also postulated that other GLUT isoforms (glucose transporters that have not yet been proposed to be regulated by hypoxia, i.e., GLUT9) might be necessary for disc cell hexose transport as has been shown to be the case for articular chondrocytes (Mobasheri et al. 2002b, 2002c; Shikhman et al. 2004, 2001a). In a recent study we tested these hypotheses by: (1) identifying expression of HIF-1α and the glucose transporter isoforms in normal human NP and AF by quantitative, real-time RT-PCR; (2) determining whether HIF-1α mRNA expression correlates with GLUT1, GLUT3, or GLUT9 mRNA expression in human NP and AF by quantitative real-time PCR; (3) localizing expression of GLUT1, GLUT3, and GLUT9 protein within regions of normal human IVD using immunohistochemistry; (4) determining whether GLUT protein expression changes with degeneration in each region of the human IVD.

Our results confirmed HIF-1α, GLUT1, GLUT3, and GLUT9 mRNA expression in NP and AF and co-expression of each GLUT isoform with HIF-1α in the NP, but not the AF. Immunohistochemistry demonstrated regional differences in GLUT expression, with the highest expression being in the NP. GLUT expression also changed as degeneration progressed. This recent study demonstrates that NP and AF cells have different GLUT expression profiles that suggest regional differences in the metabolic nature of the human IVD and that this environment changes during degeneration.

The published literature suggests that GLUT1 and GLUT3 are regulated by hypoxia (Badr et al. 1999; Vannucci et al. 1998; Zhang et al. 1999) but there is no evidence for hypoxic regulation of GLUT9. Thus, it is reasonable to suggest that upregulation of GLUT1 and GLUT3 in early disc degeneration may be a metabolic adaptation to the hypoxic conditions. Many studies have shown that GLUT3 is expressed in the brain and is important for glucose delivery to neurons when glucose concentrations are low (i.e., in hypoxic and ischemic conditions) (Vannucci et al. 1998, 1996). Upregulation of these proteins is likely to result from the cells' attempt to increase their uptake of glucose, because of decreased diffusion of glucose through the degenerate matrix, increased cellular demands from glycolytic pathways, or stimulation by catabolic, proinflammatory cytokines such as IL-1β (Phillips et al. 2005b; Richardson et al. 2003; Shikhman et al. 2001a). In summary,

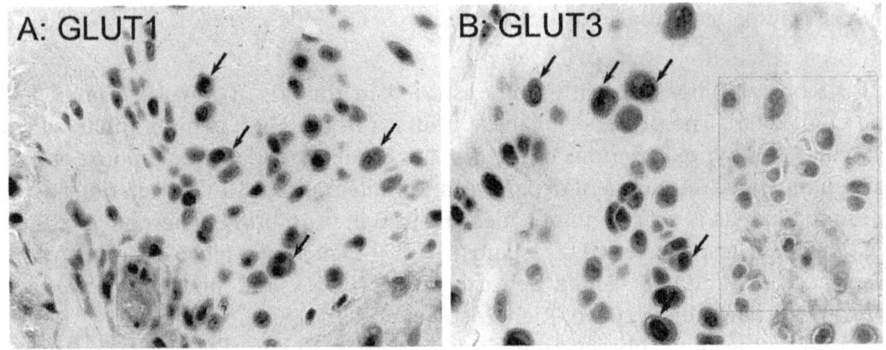

Fig. 26 Expression of GLUT1 and GLUT3 in mouse intervertebral disc chondrocytes (original magnification ×200). The immunostaining for the GLUT1 and GLUT3 proteins is indicated by the *red* substrate and the cell nuclei were counterstained with hematoxylin. The absence of GLUT3 staining in the cells in the *boxed area* in **B** suggests regional heterogeneity in GLUT isoform expression in the intervertebral disc which may be accounted for by differences in disc cell metabolism

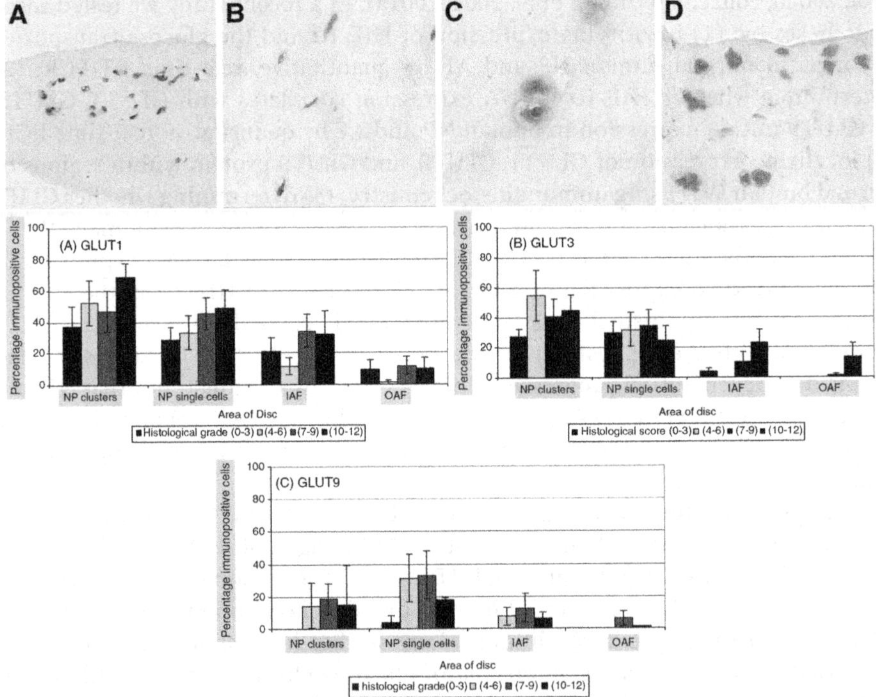

Fig. 27 Immunohistochemical localization of (**A**) HIF-1α; (**B**) GLUT1; (**C**) GLUT3; and (**D**) GLUT9 in paraffin-embedded human intervertebral disc. E Percentage of human intervertebral disc cells immunopositive for (**A**) GLUT1; (**B**) GLUT3; and (**C**) GLUT9 in each region of the human intervertebral disc at varying stages of degeneration. These glucose transporters are preferentially expressed in the nucleus pulposus region of the human intervertebral disc compared to the inner and outer annulus fibrosus

our work on human intervertebral disc cells has shown that GLUT1, GLUT3, and GLUT9 are also present in these cells (Fig. 26). We also have immunohistochemical data to confirm GLUT1 and GLUT3 expression in mouse intervertebral discs (see Fig. 27), suggesting that GLUT1, GLUT3, and GLUT9 are present in the disc along with HIF-1α and that glucose transport and metabolism are hypoxia-regulated processes in the avascular regions of the human intervertebral disc. Degeneration of the intervertebral disc will affect the expression of glucose transporters and the metabolism of its resident cells.

10
Glucose Transporter Expression and Regulation in Embryonic and Mesenchymal Stem Cells

Stem cells have the remarkable potential to develop into many different cell types in the body. Serving as a sort of repair system for the body, they can theoretically divide without limit to replenish other cells as long as the person or animal is still alive. When a stem cell divides, each new cell has the potential to either remain a stem cell or become another type of cell with a more specialized function, such as a muscle cell, a red blood cell, or a brain cell.

Stem cells are multipotential cells found in all multicellular organisms. They retain the ability to renew themselves through mitotic cell division and can differentiate into a diverse range of specialized cell types. There are two main types of mammalian stem cell: embryonic stem cells that are found in blastocysts, and adult stem cells that are found in adult tissues. In the developing embryo, stem cells differentiate into specialized embryonic tissues which give rise to different tissue and organs. Embryonic stem cells are derived from the blastocyst or earlier morula stage embryos. In humans a blastocyst is an early stage embryo approximately 4–5 days old and consists of 50–150 cells. Embryonic stem cells are pluripotent and give rise during development to all derivatives of the three primary germ layers: ectoderm, endoderm, and mesoderm. In other words, they can develop into each of the more than 200 cell types of the adult body when given sufficient and necessary stimulation for a specific cell type. In adult organisms, stem cells and progenitor cells act as a repair and regeneration system for the body, replenishing specialized cells that are dead, dying, or otherwise degenerated. Adult stem cells also maintain the normal turnover of organs with a high intrinsic regenerative capacity including blood, skin, and intestinal epithelium. Adult stem cells can be found in children as well as adults and are generally unipotent or multipotent. Pluripotent adult stem cells are very rare and are generally found in small numbers. However, they can be found in a number of tissues including umbilical cord blood. The best studied adult stem cells are multipotent and are generally referred to by their tissue origin (i.e., hematopoietic stem cells that differentiate into erythrocytes, white blood cells, platelets, etc. and bone marrow stromal cells also known as mesenchymal stem cells that have the capacity to generate bone, cartilage, fat, and fibrous connective tissues by differentiating into chondrocytes, tenocytes, adipocytes, myocytes). The hierarchy of stem cells is illustrated in Fig. 28. The differentiation pathways involved in the specialization of hematopoietic and mesenchymal stem cells are depicted in Figs. 29 and 30, respectively.

A great deal of recent research on embryonic and adult stem cells has focused on clarifying their phenotype and establishing reliable markers for their identification and fractionation. Ongoing research is aimed at understanding the signals that govern their capacity to divide or self-renew indefinitely and their differentiation potential. Some of this work has focused on understanding their basic metabolism and in this final section we review some of the recently published work on glucose transport and glucose transporters in embryonic and mesenchymal stem cells.

Human embryonic stem cells have unique features including unlimited growth capacity, expression of specific cell surface markers, normal karyotypes, and an ability to differentiate. Lee and co-workers (Lee et al. 2005) compared the phenotypic characteristics of three human embryonic stem cell lines and found significant differences in the expression levels of tissue-specific markers such as renin, kallikrein, beta- and delta-globin, albumin, alpha1-antitrypsin (alpha1-AT), and GLUT2. These embryonic stem cell lines also differed in their basic proliferative

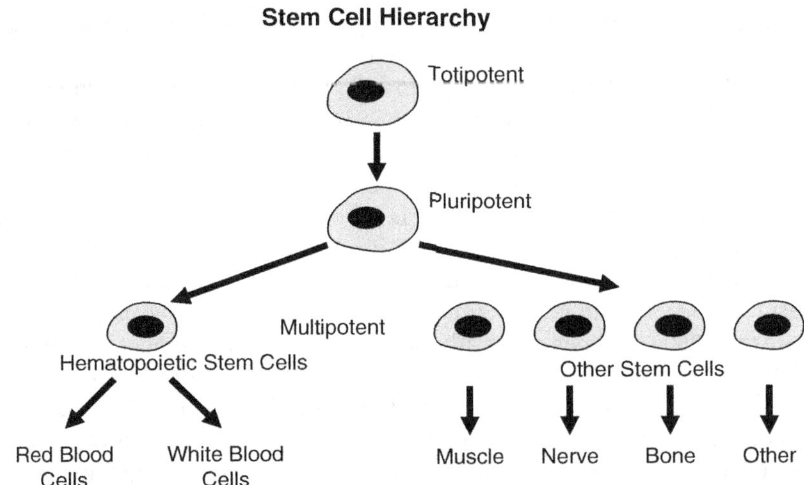

Fig. 28 The hierarchy of stem cells. The potency of stem cells specifies the differentiation potential of stem cells. Totipotent stem cells are produced from the fusion of an egg and sperm cell. Cells produced by the first few divisions of the fertilized egg are also totipotent. These cells can differentiate into embryonic and extra-embryonic cell types. Pluripotent stem cells are the descendants of totipotent cells and can differentiate into cells derived from any of the three germ layers. Multipotent stem cells can produce only cells of a closely related family of cells (e.g., hematopoietic stem cells differentiate into red blood cells, white blood cells, platelets, etc., and other stem cells which include mesenchymal stem cells). Unipotent cells only have the capacity to produce one cell type, but have the property of self-renewal which distinguishes them from nonstem cells (e.g., muscle stem cells)

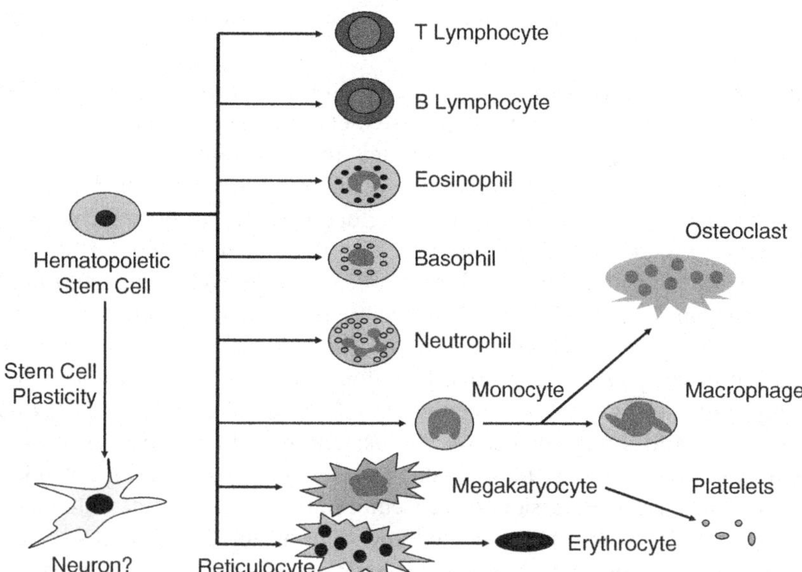

Fig. 29 Differentiation of hematopoietic stem cells into various specialized cell types. Hematopoietic stem cells have the capacity to differentiate into B and T lymphocytes, neutrophils, eosinophils, basophils, monocytes, macrophages, and megakaryocytes. Hematopoietic stem cells can also give rise to reticulocytes that form functional erythrocytes. Monocytes and macrophages can differentiate into bone resorbing osteoclasts. Some reports suggest that hematopoietic stem cells may also have the capacity to differentiate into neurons (a process called stem cell plasticity)

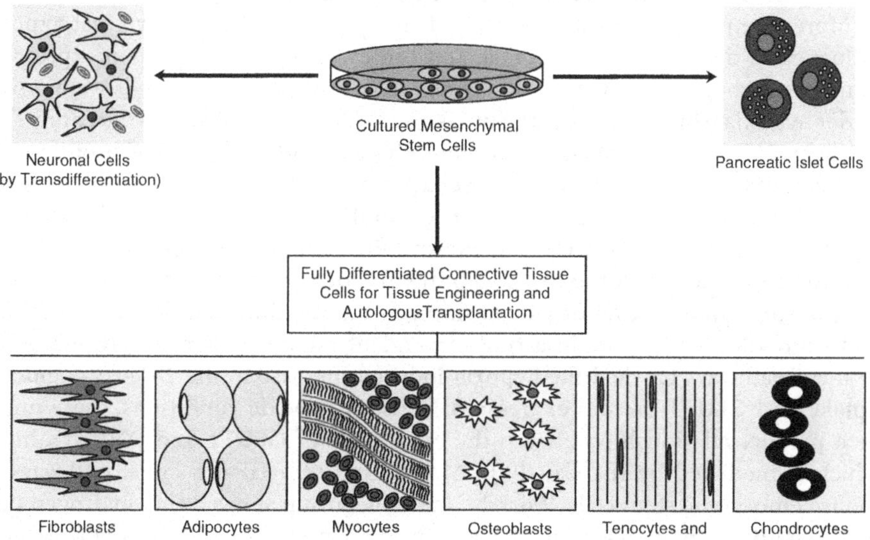

Fig. 30 Differentiation of cultured mesenchymal stem cells. Mesenchymal stem cells have the capacity to differentiate into connective tissue and musculoskeletal cells for tissue engineering, autologous implantation/transplantation, and regenerative medicine. This process involves commitment, lineage progression, differentiation, and maturation

activity. The differences they observed in GLUT2 expression may underline important differences in their basal glucose metabolism (Lee et al. 2005).

Abnormally high glucose levels may play an important role in early embryo development and function. Previous work has already established that mouse blastocysts express at least six facilitative glucose transporter isoforms (GLUT1, GLUT2, GLUT3, GLUT4, GLUT8, and GLUT9) (Pantaleon and Kaye 1998; Watson and Pessin 2001). Studies by Tonack and colleagues (Tonack et al. 2006) used the mouse embryonic stem cell line D3 and spontaneously differentiating embryoid bodies to investigate GLUT expression and the influence of glucose during differentiation of early embryonic cells. They found that both embryonic stem cells and embryoid bodies (2d–20d) expressed GLUT1, GLUT3, and GLUT8, whereas the isoforms 2 and 4 were detectable exclusively in embryoid bodies. Analogous to trophoblast cells in mouse blastocysts the outer cell layer of endoderm-like cells showed a high GLUT3 expression in early embryoid bodies. In 20-day-old embryoid bodies no GLUT3 protein was detected and only minor GLUT3 mRNA amounts were detectable. A minimal glucose concentration of 5 mM applied during 2 and 8 days of embryoid body culture resulted in upregulation of GLUT4. The authors concluded that GLUT expression in embryonic stem cells depends on the differentiation state and that GLUT expression is modulated by glucose concentration (Tonack et al. 2006). Although this idea is not novel the findings support the long-held notion that substrate concentration is an important regulator of GLUT isoform expression. The developmental and glucose-dependent regulation of GLUT isoforms suggests that glucose and its transporters play important functional roles in embryonic stem cell differentiation and embryonic development.

More recent regulatory studies carried out by Lee et al. have shown that hypoxia influences 2-deoxy-D-glucose uptake and cell cycle regulatory protein expression in mouse embryonic stem cells (Lee et al. 2007). Embryonic stem cells were grown under hypoxia which increased the uptake of 2-deoxy-D-glucose and the expression of the hypoxia-responsive GLUT1 protein. Hypoxia did not affect the undifferentiated state of ES cells and cell viability over a period of 48 h. The authors also observed a significant increase in ^{3}H thymidine incorporation at 12 h after hypoxic exposure. Hypoxia also increased the Ca^{2+} uptake and PKC beta (I), epsilon, and zeta translocation from the cytosol to the membrane fraction. Moreover, hypoxia increased the phosphorylation level of p44/42 mitogen-activated protein kinases (MAPKs) and expression of HIF-1α in a time-dependent manner. Interestingly, inhibition of these pathways blocked the hypoxia-induced increase in the 2-deoxy-D-glucose uptake and GLUT1 protein expression. Thus, in hypoxic conditions, there might be a parallel relationship between the expression of GLUT1 and DNA synthesis, which is mediated by the Ca^{2+}/PKC, MAPK, and the HIF-1α signal pathways in mouse embryonic stem cells. Work by the same group showed that hydrogen peroxide significantly increases the level of 2-deoxy-D-glucose uptake in a time- and concentration-dependent manner. Furthermore, hydrogen peroxide was found to cause a significant increase in the mRNA and protein level of GLUT1. The authors employed pharmacological tools to demonstrate that hydrogen peroxide increases the 2-deoxy-D-glucose uptake via MAPKs, cytosolic phospholipase A(2), and NF-κ B signaling pathways in mouse embryonic stem cells (Na et al. 2007).

GLUT1 is the major glucose transporter expressed in the fertilized egg and it is abundantly produced in the preimplantation embryo. Haploinsufficiency occurs when a diploid organism only has a single functional copy of a gene. In the case of GLUT1 haploinsufficiency the result is GLUT1 deficiency syndrome in humans. However, the embryo of the GLUT1 haploinsufficient human appears unaffected. In an effort to understand the mechanisms responsible for this apparent anomaly, Heilig and co-workers produced heterozygous GLUT1 knockout murine embryonic stem cells (GT1$^{+/-}$) to study the role of GLUT1 deficiency in their growth, glucose metabolism, and survival in response to hypoxic stress (Heilig et al. 2003). Their insightful approach to this problem revealed that GT1($^{-/-}$) cells are nonviable. However, both the GT1($^{+/+}$) and GT1($^{+/-}$) embryonic stem cells expressed the GLUT1 and GLUT3 high-affinity, facilitative glucose transporters. GT1($^{+/-}$) cells demonstrated 49±4% reduction of GLUT1 mRNA. This induced a post-transcriptional, GLUT1 compensatory response resulting in 24±4% reduction of GLUT1 protein. GLUT3 expression levels were unchanged. GLUT8 and GLUT12 were also expressed and unchanged in GT1($^{+/-}$). Stimulation of glycolysis by inhibition of oxidative phosphorylation using azide was impaired by 44% in GT1($^{+/-}$), with impaired upregulation of GLUT1 protein. Hypoxia for up to 4 h led to 201% more apoptosis in GT1($^{+/-}$) than in GT1($^{+/+}$) controls. Caspase-3 activity was 76% higher in GT1($^{+/-}$) versus GT1($^{+/+}$) after 2 h of hypoxia. Heterozygous knockout of GLUT1 led to a partial GLUT1 compensatory response protecting nonstressed cells. However, inhibition of oxidative phosphorylation and hypoxia both exposed their increased susceptibility to these environmental stresses (Heilig et al. 2003).

The tissue distribution of glucose transporters is not constant throughout embryonic development (Santalucia et al. 1992). High levels of GLUT1 and GLUT3 are present in a wide range of embryonic and fetal tissues, with expression of these transporters greatly decreased after birth in many of these cell types. Abundant levels of the GLUT1 and GLUT3 proteins are also present in preimplantation mouse embryos, since glucose is the main substrate consumed (Pantaleon and Kaye 1998; Pantaleon et al. 2001). During the early period of organ formation (i.e., brain, heart, skeletal muscle, and kidney) GLUT1 is responsible for glucose supply to the dividing and differentiating cells (Matsumoto et al. 1995; Santalucia et al. 1992). There is evidence that GLUT8 (the novel IGF-I and insulin-regulated glucose transporter) and GLUT9 both play a role in preimplantation development (Carayannopoulos et al. 2000, 2004; Wyman et al. 2003).

Our recent work and work by Shikhman and co-workers have shown that GLUT1 and GLUT9 are present in chondrocytes derived from fully developed human articular cartilage (Mobasheri et al. 2002b; Richardson et al. 2003; Shikhman et al. 2001a). However, very little research work has been done on the expression of these two glucose transporter isoforms during cartilage development. Until recently it was not known if GLUT1 and GLUT9 are present in articular cartilage and musculo-skeletal tissues of postimplantation embryos. Thus, in a recent paper we tested the hypothesis that glucose transporters are important for chondrogenesis during embryonic development and that GLUT1 and GLUT9 may be important in this process. Accordingly, we reported on the developmental expression of GLUT1 and GLUT9 in embryonic, juvenile, and mature ovine cartilage and provided evidence for the presence of these glucose transporter proteins in embryonic ovine chondroblasts and mature ovine articular chondrocytes (Mobasheri et al. 2005a) (Figs. 31, 32, 33, and 34). The major findings of this investigation suggested that in addition to GLUT3, GLUT4, and GLUT12, developing and mature mammalian 'chondroblasts' also express GLUT1 and the recently described GLUT9 glucose transporter. However, unlike GLUT4 and GLUT12, GLUT9 and GLUT1 appear to be expressed during development and in maturity in articular cartilage and may be important for the maintenance and homeostasis of fully developed cartilage. The presence of GLUT1 and GLUT9 in embryonic ovine chondroblasts supports a critical role for these glucose transporters and possibly others (GLUT3 and GLUT12) in ovine cartilage development. The GLUT isoforms implicated in embryonic chondroblasts may be involved in transporting hexose and pentose sugars (i.e., glucose and fructose respectively), sulfated sugars, and possibly also dehydroascorbic acid (vitamin C), which is involved in hydroxyproline and hydroxylysine synthesis for incorporation into newly synthesized cartilage matrix collagens. These glucose transporters are important for the maintenance and homeostasis of fully developed cartilage matrix. Chondrocytes also possess the capacity for gluconeogenesis via the

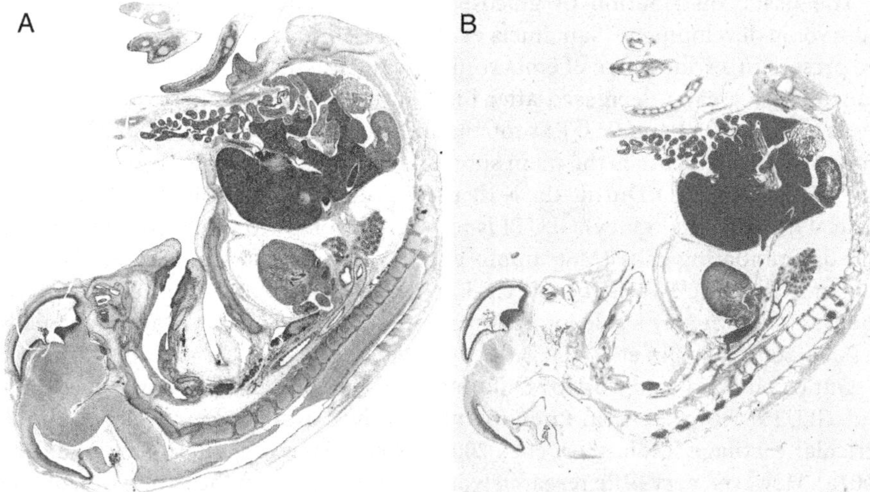

Fig. 31 **A** Expression of GLUT1 in the ovine embryo. Immunohistochemistry reveals abundant levels of GLUT1 expression in many developing embryonic tissues in the sheep embryo including brain, liver, kidney, intestines, and spine. **B** An H&E-stained section

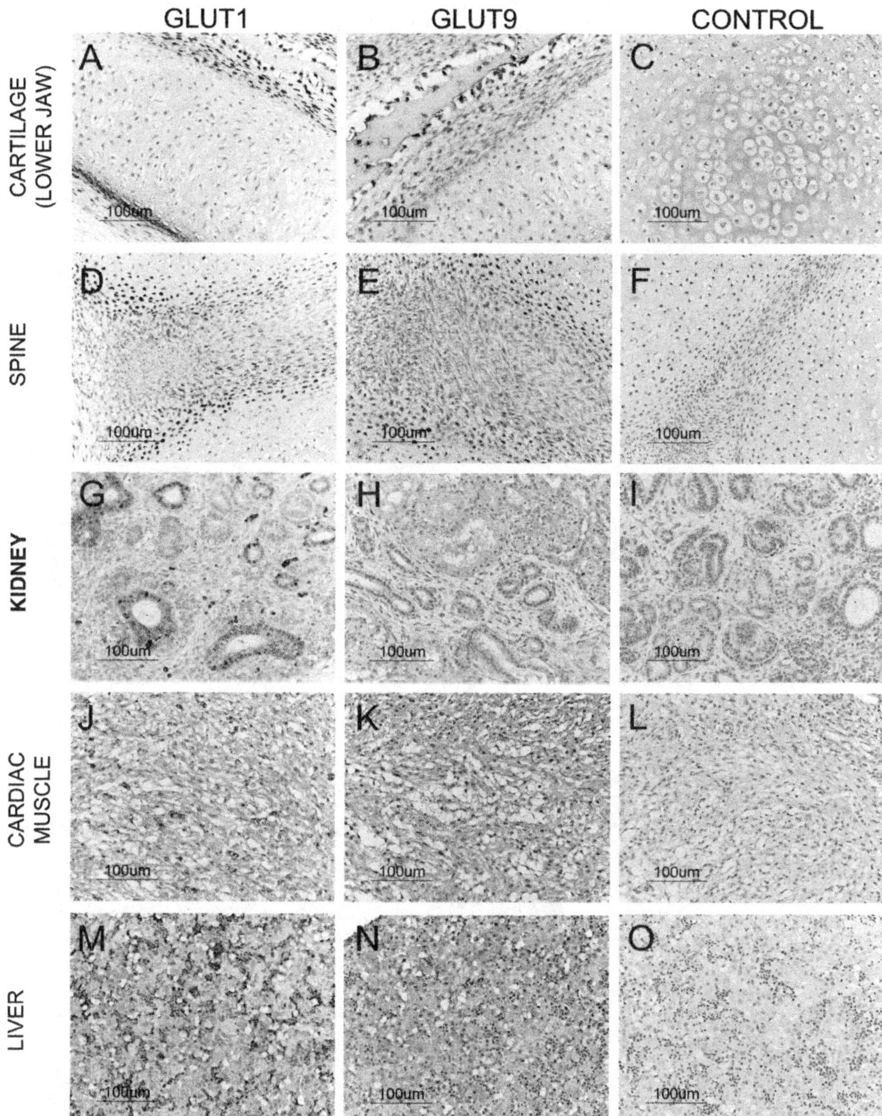

Fig. 32 Immunoperoxidase localization of GLUT1 and GLUT9 in selected tissues of E32-E36 ovine embryos. The images shown are representative sections taken from two ovine fetuses at this developmental stage. Expression of GLUT1 and GLUT9 is shown in the cartilage and mineralizing bone of the lower mandible (**A** and **D**) and in the spine (**B** and **E**). High levels of GLUT1 and GLUT9 expression were also detected in the heart and liver (**J, K, M,** and **N**). GLUT1 immunostaining in the developing kidney was observed in basolateral membranes of renal tubules and erythrocytes (**G**). GLUT9 immunostaining in the developing kidney was weak and restricted to nephron structures resembling proximal tubules (**H**). The negative controls demonstrate that nonspecific immunostaining did not occur when the primary antibody was omitted from the immunohistochemical protocol. *Bars* represent 100 μm and nuclei were counterstained with hematoxylin. (Reproduced from Mobasheri et al. 2005a with copyright permission of Elsevier Science)

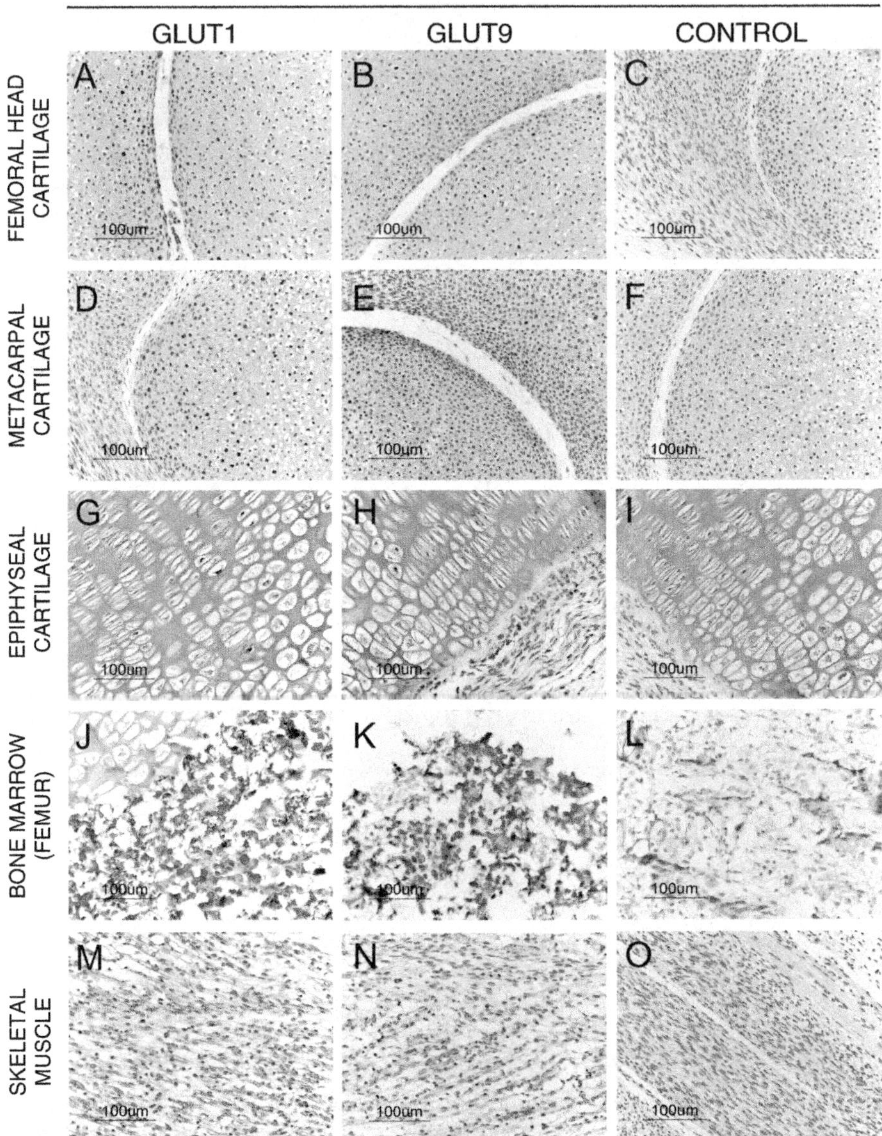

Fig. 33 Immunoperoxidase localization of GLUT1 and GLUT9 in connective tissues of E42-E45 ovine embryos. The images shown are representative sections taken from four ovine fetuses at this developmental stage. Expression of GLUT1 and GLUT9 is shown in femoral head, metacarpal and epiphyseal cartilage and mineralizing bone of the femur (**A, B, D, E, G, H, J,** and **K**). Expression of GLUT1 and GLUT9 is also evident in skeletal muscle. The negative controls demonstrate that nonspecific immunostaining did not occur when the slides were incubated with nonimmune rabbit serum (**C, F, I, L**, and **O**). Nuclei were counterstained with hematoxylin and *bars* represent 100 μm. (Reproduced from Mobasheri et al. 2005a with copyright permission of Elsevier Science)

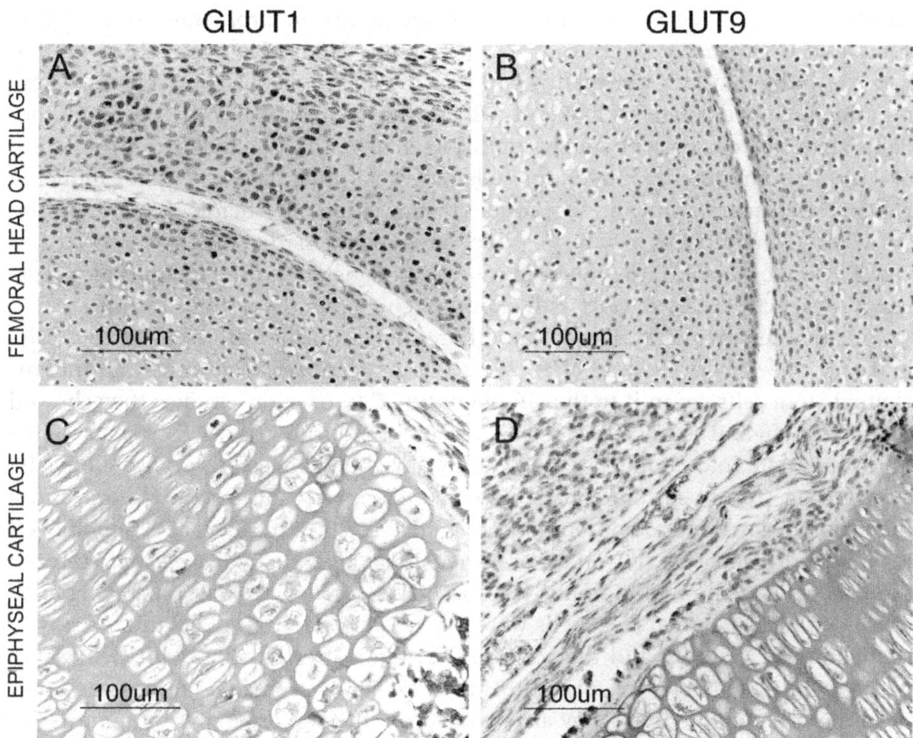

Fig. 34 High-magnification immunohistochemical micrographs showing the cellular localization of GLUT1 and GLUT9 in articular cartilage and epiphyseal cartilage from joints of E42 to E45 ovine embryos. Plasma membrane and cytoplasmic GLUT1 staining is evident in **A** and **C**. Cytoplasmic and nuclear membrane localization is seen with anti-GLUT9 antibodies (**B** and **D**)

glyoxylate pathway and some of the GLUTs may be involved in the glucose fluxes that may occur in physiological scenarios where these pathways are functional.

Mesenchymal stem cells (also known as bone marrow stromal cells) give rise to a variety of cell types: bone cells (osteoblasts, osteocytes), cartilage cells (chondrocytes), fat cells (adipocytes), and other kinds of connective tissue cells such as those in tendons and ligaments. Very few original studies have explored the expression and regulation of GLUT proteins in mesenchymal stem cells. Recent work on umbilical cord blood has suggested that this unique blood source may be used as a potential source of various kinds of stem cells, including hematopoietic stem cells, mesenchymal stem cells, and endothelial progenitor cells for a variety of cell therapies. Nagano and co-workers recently reported a novel method for isolating endothelial progenitor cells from umbilical cord blood by a combination of negative immunoselection and cell culture techniques (Nagano et al. 2007). They divided endothelial progenitor cells into two subpopulations according to their aldehyde dehydrogenase (ALDH) activity. They observed that endothelial progenitor cells with low

ALDH activity (Alde-Low) possess a greater ability to proliferate and migrate compared to those with high ALDH activity (Alde-High). They also observed that hypoxia-inducible factor-regulated proteins are upregulated in these cells. These proteins included VEGF and GLUT1, which both increased in Alde-Low endothelial progenitor cells under hypoxic conditions. They proposed that the introduction of Alde-Low endothelial progenitor cells may be a potential strategy for inducing rapid neovascularization and subsequent regeneration of ischemic tissues (Nagano et al. 2007). Therefore, it may be possible that hypoxia markers such as HIF-1α and GLUT1 (and other GLUT markers) may be used as metabolic and phenotypic markers of mesenchymal stem cells with chondrogenic potential in future studies.

Since mesenchymal stem cells can differentiate into chondrocyte-like cells, Risbud and co-workers (Risbud et al. 2004a) asked the basic question: 'Can mesenchymal stem cells commit to the NP phenotype?' In previous studies they showed that mesenchymal stem cells such as those obtained from marrow stroma are able to differentiate into chondrocyte-like NP cells when exposed to a physiologically appropriate culture system consisting of hypoxia, relevant growth factors, and a three-dimensional microenvironment (Risbud et al. 2003, 2004b). They immobilized rat mesenchymal stem cells in three-dimensional alginate hydrogels and cultured the gels in a medium containing TGF-β1 under hypoxia (2% O_2) and normoxia (20% O_2). Mesenchymal stem cells were then examined by confocal microscopy to evaluate their viability and their metabolic status after labeling with Celltracker green, a thiol-sensitive dye, and Mitotracker red, a dye sensitive to the mitochondrial membrane potential. They also carried out flow cytometry, semiquantitative RT-PCR, and Western blot analysis to evaluate phenotypic and biosynthetic activities and the signaling pathways involved in the differentiation process. They found that under hypoxic conditions, mesenchymal stem cells formed large aggregates and exhibited positive Celltracker and Mitotracker signals. GLUT3, MMP2, collagen type II and type XI, and aggrecan mRNA and protein expression were all upregulated in hypoxic conditions. Hypoxia also increased the expression of β3 and α2 integrin. Treatment with TGF-β1 increased MAPK activity, Sox-9, aggrecan, and collagen type II gene expression. Their results indicate that hypoxia and TGF-β1 drive mesenchymal stem cell differentiation towards a phenotype consistent with that of the chondrocyte-like cell found in the NP of the intervertebral disc. Our recent observation of GLUT3 expression (along with GLUT1 and GLUT9) in human intervertebral disc also confirms the results obtained by Risbud and co-workers and suggests that chondrocyte-like cells in the NP of the intervertebral disc express hypoxia-responsive glucose transporters (Risbud et al. 2004a) (Figs. 26 and 27).

11
Concluding Remarks

Understanding glucose transport and metabolism in chondrocytes is a key to understanding the processes of chondrogenesis, skeletal development, and cartilage degradation in OA. Our studies have allowed us to construct a new 'metabolic' and phenotypic signature for these cells based on the knowledge that articular cartilage is an avascular, noninsulin-sensitive, and glycolytic tissue that needs to utilize glucose and that oxygen- and nutrient-deprived chondrocytes express three key phenotypic markers: hypoxia-inducible factor alpha (HIF-1α) (Pfander et al. 2003; Rajpurohit et al. 2002; Schipani et al. 2001), glucose transporter 1 (GLUT1), and glucose transporter 3 (GLUT3) (Mobasheri et al. 2002b; Richardson et al. 2003). The presence of these proteins in chondrocytes suggests that chondrocytes in avascular and hypoxic regions of cartilage are capable of adaptive responses to low oxygen tensions and low glucose concentrations. We propose that the metabolic adjustment of chondrocytes to a low-oxygen, low-glucose environment involves induction of the HIF-1α oxygen-sensing transcription factor, which activates target genes involved in increasing glucose uptake (i.e., the hypoxia-responsive GLUT1 and GLUT3 facilitative glucose transporters) (Vannucci et al. 1998, 1996) and those promoting anaerobic glycolysis [i.e., lactate dehydrogenase (LDH) and phosphofructokinase (PFK)] (Semenza et al. 1997, 1996, 1994). Based on our recent studies of glucose transporters in articular chondrocytes we propose that hypoxia-responsive glucose transporters GLUT1 and GLUT3 function as the primary receptors of 'extracellular glucose-sensing' apparatus and that their expression is regulated by activation of the HIF-1α transcription factor in hypoxia. SGLT, GLUT4, and GLUT2 are not expressed in chondrocytes and are unlikely to be involved in glucose sensing in these cells. It is not known if other SGLT isoforms (i.e., SGLT2 and SGLT3) are present in cartilage. Clearly more experiments are required to identify other components of the glucose-sensing mechanism in chondrocytes. However, identification of some of the putative signaling components is already underway. Recent work by Shikhman and co-workers has shown that distinct signaling pathways regulate glucose transport in human chondrocytes in response to anabolic (TGF-β) and catabolic (IL-1β) mediators (Shikhman et al. 2004, 2001a). IL-1β regulation of glucose transport in human chondrocytes depends on protein kinase C (PKC) and ERK/p38 signal transduction pathways, and does not require phosphoinositide 3-kinase, extracellular signal-related kinase, or c-Jun N-terminal kinase activation (Shikhman et al. 2001a). The role of hexokinase in 'cytoplasmic' or 'intracellular glucose sensing' and metabolic regulation in chondrocytes is not known and is worthy of further investigation. It is possible that other GLUT isoforms that are not regulated by hypoxia (i.e., the recently described GLUT9 isoform) might also be necessary for glucose sensing in the ECM of cartilage (Mobasheri et al. 2002b, 2002c; Richardson et al. 2003; Shikhman et al. 2004). Our preliminary data also suggest that GLUT9 is expressed in the intervertebral disc (lower levels than GLUT 1 and GLUT3) and its expression shows no correlation with HIF-1α (Richardson et al., unpublished observations; Fig. 27).

In this monograph we have reviewed the available literature on glucose transporter expression and regulation in chondrocytes. It is clear that much more work is needed to understand the metabolism of chondrocyte precursors, especially mesenchymal stem cells. We have proposed that some GLUTs may be intimately involved in glucose sensing in chondrocytes and attempted to link these studies to the work currently being done on hypoxic conditioning in cartilage. We hope that our work will stimulate further interest, research, and debate in this area of basic research related to orthopedics and rheumatology, which will undoubtedly be important for understanding cartilage metabolism in health and disease.

Acknowledgements

A. Mobasheri would like to acknowledge the financial support of the Wellcome Trust, BBSRC, EPSRC and the Waltham Centre for Pet Nutrition (a division of Mars-Masterfoods, UK). A. Mobasheri would also like to thank all the former dissertation, vacation, Master's and PhD students who have worked in his laboratory. Special thanks go to my former doctoral thesis advisors in Oxford and Edinburgh for exceptional professionalism, kindness, support and encouragement over the last 10 years and more. They have provided invaluable help, advice and encouragement. I am also grateful for all their early and recent political efforts to facilitate my scientific career progression.

References

Adams CS, Horton WE, Jr (1998) Chondrocyte apoptosis increases with age in the articular cartilage of adult animals. Anat Rec 250:418–425

Aigner T, Fundel K, Saas J, Gebhard PM, Haag J, et al. (2006a) Large-scale gene expression profiling reveals major pathogenetic pathways of cartilage degeneration in osteoarthritis. Arthritis Rheum 54:3533–3544

Aigner T, Sachse A, Gebhard PM, Roach HI (2006b) Osteoarthritis: pathobiology-targets and ways for therapeutic intervention. Adv Drug Deliv Rev 58:128–149

Airley RE, Mobasheri A (2007) Hypoxic regulation of glucose transport, anaerobic metabolism and angiogenesis in cancer: novel pathways and targets for anticancer therapeutics. Chemotherapy 53:233–256

Alpert E, Gruzman A, Totary H, Kaiser N, Reich R, Sasson S (2002) A natural protective mechanism against hyperglycaemia in vascular endothelial and smooth-muscle cells: role of glucose and 12-hydroxyeicosatetraenoic acid. Biochem J 362:413–422

Alvarez E, Martinez MD, Roncero I, Chowen JA, Garcia-Cuartero B, et al. (2005) The expression of GLP-1 receptor mRNA and protein allows the effect of GLP-1 on glucose metabolism in the human hypothalamus and brainstem. J Neurochem 92:798–806

Archer CW, Francis-West P (2003) The chondrocyte. Int J Biochem Cell Biol 35:401–404

Arluison M, Quignon M, Nguyen P, Thorens B, Leloup C, Penicaud L (2004a) Distribution and anatomical localization of the glucose transporter 2 (GLUT2) in the adult rat brain-- an immunohistochemical study. J Chem Neuroanat 28:117–136

Arluison M, Quignon M, Thorens B, Leloup C, Penicaud L (2004b) Immunocytochemical localization of the glucose transporter 2 (GLUT2) in the adult rat brain. II. Electron microscopic study. J Chem Neuroanat 28:137–146

Arnott G, Coghill G, McArdle HJ, Hundal HS (1994) Immunolocalization of GLUT1 and GLUT3 glucose transporters in human placenta. Biochem Soc Trans 22:272S

Ashcroft FM, Gribble FM (1998) Correlating structure and function in ATP-sensitive K+ channels. Trends Neurosci 21:288–294

Babenko AP, Aguilar-Bryan L, Bryan J (1998) A view of sur/KIR6.X, KATP channels. Annu Rev Physiol 60:667–687

Badr GA, Zhang JZ, Tang J, Kern TS, Ismail-Beigi F (1999) Glut1 and glut3 expression, but not capillary density, is increased by cobalt chloride in rat cerebrum and retina. Brain Res Mol Brain Res 64:24–33

Barracchini A, Franceschini N, Amicosante G, Oratore A, Minisola G, et al. (1998) Can non-steroidal anti-inflammatory drugs act as metalloproteinase modulators? An in-vitro study of inhibition of collagenase activity. J Pharm Pharmacol 50:1417–1423

Barracchini A, Franceschini N, Minisola G, Pantaleoni GC, Di Giulio AD, et al. (1999) Meloxicam and indomethacin activity on human matrix metalloproteinases in synovial fluid. Ann N Y Acad Sci 878:665–666

Bartels EM, Fairbank JC, Winlove CP, Urban JP (1998) Oxygen and lactate concentrations measured in vivo in the intervertebral discs of patients with scoliosis and back pain. Spine 23:1–7; discussion 8

Bayliss MT, Ali SY (1978) Age-related changes in the composition and structure of human articular-cartilage proteoglycans. Biochem J 176:683–693

Behrooz A, Ismail-Beigi F (1997) Dual control of glut1 glucose transporter gene expression by hypoxia and by inhibition of oxidative phosphorylation. J Biol Chem 272:5555–5562

Behrooz A, Ismail-Beigi F (1999) Stimulation of glucose transport by hypoxia: signals and mechanisms. News Physiol Sci 14:105–110

Bell GI, Kayano T, Buse JB, Burant CF, Takeda J, et al. (1990) Molecular biology of mammalian glucose transporters. Diabetes Care 13:198–208

Bell GI, Murray JC, Nakamura Y, Kayano T, Eddy RL, et al. (1989) Polymorphic human insulin-responsive glucose-transporter gene on chromosome 17p13. Diabetes 38:1072–1075

Benya PD, Shaffer JD (1982) Dedifferentiated chondrocytes reexpress the differentiated collagen phenotype when cultured in agarose gels. Cell 30:215–224

Bernick S, Cailliet R (1982) Vertebral end-plate changes with aging of human vertebrae. Spine 7:97–102

Bibby SR, Jones DA, Lee RB, Yu J, Urban JPG (2001) The pathophysiology of the intervertebral disc. Joint Bone Spine 68:537–542

Bird TA, Davies A, Baldwin SA, Saklatvala J (1990) Interleukin 1 stimulates hexose transport in fibroblasts by increasing the expression of glucose transporters. J Biol Chem 265:13578–13583

Birnbaum MJ (1989) Identification of a novel gene encoding an insulin-responsive glucose transporter protein. Cell 57:305–315

Bisson LF, Coons DM, Kruckeberg AL, Lewis DA (1993) Yeast sugar transporters. Crit Rev Biochem Mol Biol 28:259–308

Blanco FJ, Guitian R, Vazquez-Martul E, de Toro FJ, Galdo F (1998) Osteoarthritis chondrocytes die by apoptosis. A possible pathway for osteoarthritis pathology. Arthritis Rheum 41:284–289

Blanco FJ, Ochs RL, Schwarz H, Lotz M (1995) Chondrocyte apoptosis induced by nitric oxide. Am J Pathol 146:75–85

Blot L, Marcelis A, Devogelaer JP, Manicourt DH (2000) Effects of diclofenac, aceclofenac and meloxicam on the metabolism of proteoglycans and hyaluronan in osteoarthritic human cartilage. Br J Pharmacol 131:1413–1421

Boehm AK, Seth M, Mayr KG, Fortier LA (2007) Hsp90 mediates insulin-like growth factor 1 and interleukin-1beta signaling in an age-dependent manner in equine articular chondrocytes. Arthritis Rheum 56:2335–2343

Bridges CH, Womack JE, Harris ED, Scrutchfield WL (1984) Considerations of copper metabolism in osteochondrosis of suckling foals. J Am Vet Med Assoc 185:173–178

Bruckner BA, Ammini CV, Otal MP, Raizada MK, Stacpoole PW (1999) Regulation of brain glucose transporters by glucose and oxygen deprivation. Metabolism 48:422–431

Buckwalter JA, Mankin HJ (1998a) Articular cartilage: degeneration and osteoarthritis, repair, regeneration, and transplantation. Instr Course Lect 47:487–504

Buckwalter JA, Mankin HJ (1998b) Articular cartilage: tissue design and chondrocyte-matrix interactions. Instr Course Lect 47:477–486

Buckwalter JA, Martin J, Mankin HJ (2000) Synovial joint degeneration and the syndrome of osteoarthritis. Instr Course Lect 49:481–489

Burant CF, Davidson NO (1994) GLUT3 glucose transporter isoform in rat testis: localization, effect of diabetes mellitus, and comparison to human testis. Am J Physiol 267: R1488–495

Burant CF, Takeda J, Brot-Laroche E, Bell GI, Davidson NO (1992) Fructose transporter in human spermatozoa and small intestine is GLUT5. J Biol Chem 267:14523–14526

Burkus JK, Ganey TM, Ogden JA (1993) Development of the cartilage canals and the secondary center of ossification in the distal chondroepiphysis of the prenatal human femur. Yale J Biol Med 66:193–202

Burstein DE, Reder I, Weiser K, Tong T, Pritsker A, Haber RS (1998) GLUT1 glucose transporter: a highly sensitive marker of malignancy in body cavity effusions. Mod Pathol 11:392–396

Carayannopoulos MO, Chi MM, Cui Y, Pingsterhaus JM, McKnight RA, et al. (2000) GLUT8 is a glucose transporter responsible for insulin-stimulated glucose uptake in the blastocyst. Proc Natl Acad Sci USA 97:7313–7318

Carayannopoulos MO, Schlein A, Wyman A, Chi M, Keembiyehetty C, Moley KH (2004) GLUT9 is differentially expressed and targeted in the preimplantation embryo. Endocrinology 145:1435–1443

Carney SL, Muir H (1988) The structure and function of cartilage proteoglycans. Physiol Rev 68:858–910

Casey A (2003) Hormonal control of metabolism: regulation of plasma glucose. Surgery (Oxford) 21:128a–d

Cawston T, Billington C, Cleaver C, Elliott S, Hui W, et al. (1999) The regulation of MMPs and TIMPs in cartilage turnover. Ann NY Acad Sci 878:120–129

Cecil DL, Johnson K, Rediske J, Lotz M, Schmidt AM, Terkeltaub R (2005) Inflammation-induced chondrocyte hypertrophy is driven by receptor for advanced glycation end products. J Immunol 175:8296–8302

Chappard D, Alexandre C, Riffat G (1986) Uncalcified cartilage resorption in human fetal cartilage canals. Tissue Cell 18:701–707

Charron MJ, Brosius FC, 3rd, Alper SL, Lodish HF (1989) A glucose transport protein expressed predominately in insulin-responsive tissues. Proc Natl Acad Sci USA 86:2535–2539

Chen FH, Herndon ME, Patel N, Hecht JT, Tuan RS, Lawler J (2007) Interaction of cartilage oligomeric matrix protein/thrombospondin 5 with aggrecan. J Biol Chem 282:24591–24598

Chikanza I, Fernandes L (2000) Novel strategies for the treatment of osteoarthritis. Expert Opin Investig Drugs 9:1499–1510

Choeiri C, Staines W, Messier C (2002) Immunohistochemical localization and quantification of glucose transporters in the mouse brain. Neuroscience 111:19–34

Christie MJ (1995) Molecular and functional diversity of K+ channels. Clin Exp Pharmacol Physiol 22:944–951

Clark AG, Rohrbaugh AL, Otterness I, Kraus VB (2002) The effects of ascorbic acid on cartilage metabolism in guinea pig articular cartilage explants. Matrix Biol 21:175–184

Clark JM (1990) The structure of vascular channels in the subchondral plate. J Anat 171:105–115

Clegg PD, Mobasheri A (2003) Chondrocyte apoptosis, inflammatory mediators and equine osteoarthritis. Vet J 166:3–4

Coetzee WA, Amarillo Y, Chiu J, Chow A, Lau D, et al. (1999) Molecular diversity of K+ channels. Ann NY Acad Sci 868:233–285

Coimbra IB, Jimenez SA, Hawkins DF, Piera-Velazquez S, Stokes DG (2004) Hypoxia inducible factor-1 alpha expression in human normal and osteoarthritic chondrocytes. Osteoarthritis Cartilage 12:336–345

Cole AA, Wezeman FH (1985) Perivascular cells in cartilage canals of the developing mouse epiphysis. Am J Anat 174:119–129

Cole AA, Wezeman FH (1987) Morphometric analysis of cartilage canals in the developing mouse epiphysis. Acta Anat (Basel) 128:93–97

Cramer T, Schipani E, Johnson RS, Swoboda B, Pfander D (2004) Expression of VEGF isoforms by epiphyseal chondrocytes during low-oxygen tension is HIF-1 alpha dependent. Osteoarthritis Cartilage 12:433–439

Crean JK, Roberts S, Jaffray DC, Eisenstein SM, Duance VC (1997) Matrix metalloproteinases in the human intervertebral disc: role in disc degeneration and scoliosis. Spine 22:2877–2884

Dart C, Standen NB (1993) Adenosine-activated potassium current in smooth muscle cells isolated from the pig coronary artery. J Physiol 471:767–786

Dart C, Standen NB (1995) Activation of ATP-dependent K+ channels by hypoxia in smooth muscle cells isolated from the pig coronary artery. J Physiol 483 (Pt 1): 29–39

Davenport-Goodall CL, Boston RC, Richardson DW (2004) Effects of insulin-like growth factor-II on the mitogenic and metabolic activities of equine articular cartilage with and without interleukin 1-beta. Am J Vet Res 65:238–244

Davidson NO, Hausman AM, Ifkovits CA, Buse JB, Gould GW, et al. (1992) Human intestinal glucose transporter expression and localization of GLUT5. Am J Physiol 262: C795–800

Davies ME, Pasqualicchio M, Henson F, HernandezVidal G (1996) Effects of copper and zinc on chondrocyte behaviour and matrix turnover. Pferdeheilkunde 12:367–370

De Ceuninck F, Caliez A, Dassencourt L, Anract P, Renard P (2004) Pharmacological disruption of insulin-like growth factor 1 binding to IGF-binding proteins restores anabolic responses in human osteoarthritic chondrocytes. Arthritis Res Ther 6:R393–403

del Rey A, Besedovsky H (1987) Interleukin 1 affects glucose homeostasis. Am J Physiol 253: R794–798

Delgado-Baeza E, Nieto-Chaguaceda A, Miralles-Flores C, Santos-Alvarez I (1992) Cartilage canal growth: experimental approach in the rat tibia. Acta Anat (Basel) 145:143–148

Denko CW, Boja B, Moskowitz RW (1994) Growth promoting peptides in osteoarthritis and diffuse idiopathic skeletal hyperostosis--insulin, insulin-like growth factor-I, growth hormone. J Rheumatol 21:1725–1730

Denko CW, Malemud CJ (1999) Metabolic disturbances and synovial joint responses in osteoarthritis. Front Biosci 4:D686–693

Dominick KL, Baker TA (2004) Racial and ethnic differences in osteoarthritis: prevalence, outcomes, and medical care. Ethn Dis 14:558–566

Dore S, Pelletier JP, DiBattista JA, Tardif G, Brazeau P, Martel-Pelletier J (1994) Human osteoarthritic chondrocytes possess an increased number of insulin-like growth factor 1 binding sites but are unresponsive to its stimulation. Possible role of IGF-1-binding proteins. Arthritis Rheum 37:253–263

Dudhia J (2005) Aggrecan, aging and assembly in articular cartilage. Cell Mol Life Sci 62:2241–2256

Edwards G, Weston AH (1995) The role of potassium channels in excitable cells. Diabetes Res Clin Pract 28 [Suppl]: S57–66

Evans ML, McCrimmon RJ, Flanagan DE, Keshavarz T, Fan X, et al. (2004) Hypothalamic ATP-sensitive K$^+$ channels play a key role in sensing hypoglycemia and triggering counterregulatory epinephrine and glucagon responses. Diabetes 53:2542–2551

Fernandes JC, Martel-Pelletier J, Pelletier JP (2002) The role of cytokines in osteoarthritis pathophysiology. Biorheology 39:237–246

Figenschau Y, Knutsen G, Shahazeydi S, Johansen O, Sveinbjornsson B (2001) Human articular chondrocytes express functional leptin receptors. Biochem Biophys Res Commun 287:190–197

Fiorito S, Magrini L, Adrey J, Mailhe D, Brouty-Boye D (2005) Inflammatory status and cartilage regenerative potential of synovial fibroblasts from patients with osteoarthritis and chondropathy. Rheumatology (Oxford) 44:164–171

Flier JS, Mueckler M, McCall AL, Lodish HF (1987a) Distribution of glucose transporter messenger RNA transcripts in tissues of rat and man. J Clin Invest 79:657–661

Flier JS, Mueckler MM, Usher P, Lodish HF (1987b) Elevated levels of glucose transport and transporter messenger RNA are induced by ras or src oncogenes. Science 235:1492–1495

Font B, Eichenberger D, Goldschmidt D, Boutillon MM, Hulmes DJ (1998) Structural requirements for fibromodulin binding to collagen and the control of type I collagen fibrillogenesis--critical roles for disulphide bonding and the C-terminal region. Eur J Biochem 254:580–587

Forsberg H, Ljungdahl PO (2001) Sensors of extracellular nutrients in Saccharomyces cerevisiae. Curr Genet 40:91–109

Forster C, Rucker M, Shakibaei M, Baumann-Wilschke I, Vormann J, Stahlmann R (1998) Effects of fluoroquinolones and magnesium deficiency in murine limb bud cultures. Arch Toxicol 72:411–419

Freemont AJ, Watkins A, Le Maitre C, Jeziorska M, Hoyland JA (2002) Current understanding of cellular and molecular events in intervertebral disc degeneration: implications for therapy. J Pathol 196:374–379

Fukumoto H, Kayano T, Buse JB, Edwards Y, Pilch PF, et al. (1989) Cloning and characterization of the major insulin-responsive glucose transporter expressed in human skeletal muscle and other insulin-responsive tissues. J Biol Chem 264:7776–7779

Fukumoto H, Seino S, Imura H, Seino Y, Eddy RL, et al. (1988) Sequence, tissue distribution, and chromosomal localization of mRNA encoding a human glucose transporter-like protein. Proc Natl Acad Sci USA 85:5434–5438

Galois L, Etienne S, Grossin L, Watrin-Pinzano A, Cournil-Henrionnet C, et al. (2004) Dose-response relationship for exercise on severity of experimental osteoarthritis in rats: a pilot study. Osteoarthritis Cartilage 12:779–786

Ganey TM, Love SM, Ogden JA (1992) Development of vascularization in the chondroepiphysis of the rabbit. J Orthop Res 10:496–510

Garvey WT, Maianu L, Zhu JH, Brechtel-Hook G, Wallace P, Baron AD (1998) Evidence for defects in the trafficking and translocation of GLUT4 glucose transporters in skeletal muscle as a cause of human insulin resistance. J Clin Invest 101:2377–2386

Gee E, Davies M, Firth E, Jeffcott L, Fennessy P, Mogg T (2007) Osteochondrosis and copper: histology of articular cartilage from foals out of copper supplemented and non-supplemented dams. Vet J 173:109–117

Geng Y, McQuillan D, Roughley PJ (2006) SLRP interaction can protect collagen fibrils from cleavage by collagenases. Matrix Biol 25:484–491

Gerstenfeld LC, Landis WJ (1991) Gene expression and extracellular matrix ultrastructure of a mineralizing chondrocyte cell culture system. J Cell Biol 112:501–513

Gilmore KS, Srinivas P, Akins DR, Hatter KL, Gilmore MS (2003) Growth, development, and gene expression in a persistent *Streptococcus gordonii* biofilm. Infect Immun 71:4759–4766

Godoy A, Ulloa V, Rodriguez F, Reinicke K, Yanez AJ, et al. (2006) Differential subcellular distribution of glucose transporters GLUT1–6 and GLUT9 in human cancer: ultrastructural localization of GLUT1 and GLUT5 in breast tumor tissues. J Cell Physiol 207:614–627

Goggs R, Vaughan-Thomas A, Clegg PD, Carter SD, Innes JF, et al. (2005) Nutraceutical therapies for degenerative joint diseases: a critical review. Crit Rev Food Sci Nutr 45:145–164

Goldring MB (1999) The role of cytokines as inflammatory mediators in osteoarthritis: lessons from animal models. Connect Tissue Res 40:1–11

Goldring MB (2000a) Osteoarthritis and cartilage: the role of cytokines. Curr Rheumatol Rep 2:459–465

Goldring MB (2000b) The role of the chondrocyte in osteoarthritis. Arthritis Rheum 43:1916–1926

Goupille P, Jayson MI, Valat JP, Freemont AJ (1998) Matrix metalloproteinases: the clue to intervertebral disc degeneration? Spine 23:1612–1626

Gribble FM, Williams L, Simpson AK, Reimann F (2003) A novel glucose-sensing mechanism contributing to glucagon-like peptide-1 secretion from the GLUTag cell line. Diabetes 52:1147–11454

Gruber HE, Hanley EN, Jr (1998) Analysis of aging and degeneration of the human intervertebral disc. Comparison of surgical specimens with normal controls. Spine 23:751–757

Guilak F (2000) The deformation behavior and viscoelastic properties of chondrocytes in articular cartilage. Biorheology 37:27–44

Guillet-Deniau I, Leturque A, Girard J (1994) Expression and cellular localization of glucose transporters (GLUT1, GLUT3, GLUT4) during differentiation of myogenic cells isolated from rat foetuses. J Cell Sci 107 (Pt 3): 487–96

Ha KY, Koh IJ, Kirpalani PA, Kim YY, Cho YK, et al. (2006) The expression of hypoxia inducible factor-1alpha and apoptosis in herniated discs. Spine 31:1309–13

Haber RS, Weinstein SP, O'Boyle E, Morgello S (1993) Tissue distribution of the human GLUT3 glucose transporter. Endocrinology 132:2538–43

Halter JB, Ward WK, Porte D, Jr., Best JD, Pfeifer MA (1985) Glucose regulation in non-insulin-dependent diabetes mellitus. Interaction between pancreatic islets and the liver. Am J Med 79:6–12

Hamrahian AH, Zhang JZ, Elkhairi FS, Prasad R, Ismail-Beigi F (1999) Activation of Glut1 glucose transporter in response to inhibition of oxidative phosphorylation. Arch Biochem Biophys 368:375–379

Haudenschild DR, McPherson JM, Tubo R, Binette F (2001) Differential expression of multiple genes during articular chondrocyte redifferentiation. Anat Rec 263:91–98

Hawker GA (2004) The quest for explanations for race/ethnic disparity in rates of use of total joint arthroplasty. J Rheumatol 31:1683–1685

Hedbom E, Antonsson P, Hjerpe A, Aeschlimann D, Paulsson M, et al. (1992) Cartilage matrix proteins. An acidic oligomeric protein (COMP) detected only in cartilage. J Biol Chem 267:6132–6136

Heilig C, Brosius F, Siu B, Concepcion L, Mortensen R, et al. (2003) Implications of glucose transporter protein type 1 (GLUT1)-haplodeficiency in embryonic stem cells for their survival in response to hypoxic stress. Am J Pathol 163:1873–1885

Henrotin Y, Kurz B, Aigner T (2005a) Oxygen and reactive oxygen species in cartilage degradation: friends or foes? Osteoarthritis Cartilage 13:643–654

Henrotin Y, Sanchez C, Balligand M (2005b) Pharmaceutical and nutraceutical management of canine osteoarthritis: present and future perspectives. Vet J 170:113–123

Henson FM, Davenport C, Butler L, Moran I, Shingleton WD, et al. (1997) Effects of insulin and insulin-like growth factors I and II on the growth of equine fetal and neonatal chondrocytes. Equine Vet J 29:441–447

Hernvann A, Aussel C, Cynober L, Moatti N, Ekindjian OG (1992) IL-1 beta, a strong mediator for glucose uptake by rheumatoid and non-rheumatoid cultured human synoviocytes. FEBS Lett 303:77–80

Hernvann A, Jaffray P, Hilliquin P, Cazalet C, Menkes CJ, Ekindjian OG (1996) Interleukin-1 beta-mediated glucose uptake by chondrocytes. Inhibition by cortisol. Osteoarthritis Cartilage 4:139–142

Hildebrand A, Romaris M, Rasmussen LM, Heinegard D, Twardzik DR, et al. (1994) Interaction of the small interstitial proteoglycans biglycan, decorin and fibromodulin with transforming growth factor beta. Biochem J 302 (Pt 2): 527–534

Hocquette JF, Abe H (2000) Facilitative glucose transporters in livestock species. Reprod Nutr Dev 40:517–533

Iatridis JC, Kumar S, Foster RJ, Weidenbaum M, Mow VC (1999a) Shear mechanical properties of human lumbar annulus fibrosus. J Orthop Res 17:732–737

Iatridis JC, Mente PL, Stokes IA, Aronsson DD, Alini M (1999b) Compression-induced changes in intervertebral disc properties in a rat tail model. Spine 24:996–1002

Ikebe T, Hirata M, Koga T (1986) Human recombinant interleukin 1-mediated suppression of glycosaminoglycan synthesis in cultured rat costal chondrocytes. Biochem Biophys Res Commun 140:386–391

Inerot S, Heinegard D, Audell L, Olsson SE (1978) Articular-cartilage proteoglycans in aging and osteoarthritis. Biochem J 169:143–156

Ishihara H, Urban JP (1999) Effects of low oxygen concentrations and metabolic inhibitors on proteoglycan and protein synthesis rates in the intervertebral disc. J Orthop Res 17:829–835

Iyer NV, Kotch LE, Agani F, Leung SW, Laughner E, et al. (1998) Cellular and developmental control of O2 homeostasis by hypoxia-inducible factor 1 alpha. Genes Dev 12:149–162

James DE, Brown R, Navarro J, Pilch PF (1988) Insulin-regulatable tissues express a unique insulin-sensitive glucose transport protein. Nature 333:183–185

Johnston M (1999) Feasting, fasting and fermenting. Glucose sensing in yeast and other cells. Trends Genet 15:29–33

Joost HG, Bell GI, Best JD, Birnbaum MJ, Charron MJ, et al. (2002) Nomenclature of the GLUT/SLC2A family of sugar/polyol transport facilitators. Am J Physiol Endocrinol Metab 282:E974–976

Joost HG, Thorens B (2001) The extended GLUT-family of sugar/polyol transport facilitators: nomenclature, sequence characteristics, and potential function of its novel members (review). Mol Membr Biol 18:247–256

Kahn SE, Porte D, Jr (1988) Islet dysfunction in non-insulin-dependent diabetes mellitus. Am J Med 85:4–8

Kayano T, Burant CF, Fukumoto H, Gould GW, Fan YS, et al. (1990) Human facilitative glucose transporters. Isolation, functional characterization, and gene localization of cDNAs encoding an isoform (GLUT5) expressed in small intestine, kidney, muscle, and adipose tissue and an unusual glucose transporter pseudogene-like sequence (GLUT6). J Biol Chem 265:13276–13282

Kayano T, Fukumoto H, Eddy RL, Fan YS, Byers MG, et al. (1988) Evidence for a family of human glucose transporter-like proteins. Sequence and gene localization of a protein expressed in fetal skeletal muscle and other tissues. J Biol Chem 263:15245–15248

Kealy RD, Lawler DF, Ballam JM, Lust G, Smith GK, et al. (1997) Five-year longitudinal study on limited food consumption and development of osteoarthritis in coxofemoral joints of dogs. J Am Vet Med Assoc 210:222–225

Kelley KM, Johnson TR, Ilan J, Moskowitz RW (1999) Glucose regulation of the IGF response system in chondrocytes: induction of an IGF-I-resistant state. Am J Physiol 276:R1164–1171

Kim DY, Taylor HW, Moore RM, Paulsen DB, Cho DY (2003) Articular chondrocyte apoptosis in equine osteoarthritis. Vet J 166:52–57

Knight DA, Weisbrode SE, Schmall LM, Reed SM, Gabel AA, et al. (1990) The effects of copper supplementation on the prevalence of cartilage lesions in foals. Equine Vet J 22:426–432

Koster JC, Permutt MA, Nichols CG (2005) Diabetes and insulin secretion: the ATP-sensitive K$^+$ channel (K ATP) connection. Diabetes 54:3065–3072

Kowalczyk DF, Gunson DE, Shoop CR, Ramberg CF, Jr (1986) The effects of natural exposure to high levels of zinc and cadmium in the immature pony as a function of age. Environ Res 40:285–300

Kuettner KE (1992) Biochemistry of articular cartilage in health and disease. Clin Biochem 25:155–163

Kumar A, Xiao YP, Laipis PJ, Fletcher BS, Frost SC (2004) Glucose deprivation enhances targeting of GLUT1 to lipid rafts in 3T3-L1 adipocytes. Am J Physiol Endocrinol Metab 286: E568–576

Kumar S, Connor JR, Dodds RA, Halsey W, Van Horn M, et al. (2001) Identification and initial characterization of 5000 expressed sequenced tags (ESTs) each from adult human normal and osteoarthritic cartilage cDNA libraries. Osteoarthritis Cartilage 9:641–653

Kume K, Satomura K, Nishisho S, Kitaoka E, Yamanouchi K, et al. (2002) Potential role of leptin in endochondral ossification. J Histochem Cytochem 50:159–169

Kurosaki M, Hori T, Takata K, Kawakami H, Hirano H (1995) Immunohistochemical localization of the glucose transporter GLUT1 in choroid plexus papillomas. Noshuyo Byori 12:69–73

Le Maitre CL, Freemont AJ, Hoyland JA (2004a) Localization of degradative enzymes and their inhibitors in the degenerate human intervertebral disc. J Pathol 204:47–54

Le Maitre CL, Hoyland JA, Freemont AJ (2004b) Studies of human intervertebral disc cell function in a constrained in vitro tissue culture system. Spine 29:1187–1195

Le Maitre CL, Freemont AJ, Hoyland JA (2005) The role of interleukin-1 in the pathogenesis of human intervertebral disc degeneration. Arthritis Res Ther 7:R732–745

Le Maitre CL, Freemont AJ, Hoyland JA (2007a) Accelerated cellular senescence in degenerate intervertebral discs: a possible role in the pathogenesis of intervertebral disc degeneration. Arthritis Res Ther 9:R45

Le Maitre CL, Pockert A, Buttle DJ, Freemont AJ, Hoyland JA (2007b) Matrix synthesis and degradation in human intervertebral disc degeneration. Biochem Soc Trans 35:652–655

Lee JB, Kim JM, Kim SJ, Park JH, Hong SH, et al. (2005) Comparative characteristics of three human embryonic stem cell lines. Mol Cells 19:31–38

Lee RB, Urban JP (1997) Evidence for a negative Pasteur effect in articular cartilage. Biochem J 321 (Pt 1):95–102

Lee RB, Urban JP (2002) Functional replacement of oxygen by other oxidants in articular cartilage. Arthritis Rheum 46:3190–3200

Lee SH, Heo JS, Han HJ (2007) Effect of hypoxia on 2-deoxyglucose uptake and cell cycle regulatory protein expression of mouse embryonic stem cells: involvement of Ca^{2+}/PKC, MAPKs and HIF-1alpha. Cell Physiol Biochem 19:269–282

Leibiger IB, Leibiger B, Berggren PO (2002) Insulin feedback action on pancreatic beta-cell function. FEBS Lett 532:1–6

Levick JR (1995) Microvascular architecture and exchange in synovial joints. Microcirculation 2:217–233

Li Q, Manolescu A, Ritzel M, Yao S, Slugoski M, Young JD, Chen XZ, Cheeseman CI (2004) Cloning and functional characterization of the human GLUT7 isoform SLC2A7 from the small intestine. Am J Physiol Gastrointest Liver Physiol 287:G236–G242.

Liote F, Orcel P (2000) Osteoarticular disorders of endocrine origin. Baillieres Best Pract Res Clin Rheumatol 14:251–276

Liu W, Burton-Wurster N, Glant TT, Tashman S, Sumner DR, et al. (2003) Spontaneous and experimental osteoarthritis in dog: similarities and differences in proteoglycan levels. J Orthop Res 21:730–737

MacDonald MH, Stover SM, Willits NH, Benton HP (1992) Regulation of matrix metabolism in equine cartilage explant cultures by interleukin 1. Am J Vet Res 53:2278–2285

Maher F, Simpson IA (1994) The GLUT3 glucose transporter is the predominant isoform in primary cultured neurons: assessment by biosynthetic and photoaffinity labelling. Biochem J 301 (Pt 2):379–384

Maher F, Vannucci SJ, Simpson IA (1994) Glucose transporter proteins in brain. FASEB J 8:1003–1011

Mahraoui L, Takeda J, Mesonero J, Chantret I, Dussaulx E, et al. (1994) Regulation of expression of the human fructose transporter (GLUT5) by cyclic AMP. Biochem J 301 (Pt 1): 169–175

Malemud CJ, Islam N, Haqqi TM (2003) Pathophysiological mechanisms in osteoarthritis lead to novel therapeutic strategies. Cells Tissues Organs 174:34–48

Manolescu A, Salas-Burgos AM, Fischbarg J, Cheeseman CI (2005) Identification of a hydrophobic residue as a key determinant of fructose transport by the facilitative hexose transporter SLC2A7 (GLUT7). J Biol Chem 280:42978–42983

Maroudas A, Stockwell RA, Nachemson A, Urban J (1975) Factors involved in the nutrition of the human lumbar intervertebral disc: cellularity and diffusion of glucose in vitro. J Anat 120:113–130

Maroudas AI (1976) Balance between swelling pressure and collagen tension in normal and degenerate cartilage. Nature 260:808–809

Martel-Pelletier J (1998) Pathophysiology of osteoarthritis. Osteoarthritis Cartilage 6:374–376

Martens G, Cai Y, Hinke S, Stange G, Van de Casteele M, Pipeleers D (2005) Nutrient sensing in pancreatic beta cells suppresses mitochondrial superoxide generation and its contribution to apoptosis. Biochem Soc Trans 33:300–301

Matsumoto K, Akazawa S, Ishibashi M, Trocino RA, Matsuo H, et al. (1995) Abundant expression of GLUT1 and GLUT3 in rat embryo during the early organogenesis period. Biochem Biophys Res Commun 209:95–102

McAlindon T, Felson DT (1997) Nutrition: risk factors for osteoarthritis. Ann Rheum Dis 56:397–400

McAlindon TE (2006) Nutraceuticals: do they work and when should we use them? Best Pract Res Clin Rheumatol 20:99–115

McGlashan SR, Jensen CG, Poole CA (2006) Localization of extracellular matrix receptors on the chondrocyte primary cilium. J Histochem Cytochem 54:1005–1014

McNulty AL, Vail TP, Kraus VB (2005) Chondrocyte transport and concentration of ascorbic acid is mediated by SVCT2. Biochim Biophys Acta 1712:212–221

Mellanen P, Minn H, Grenman R, Harkonen P (1994) Expression of glucose transporters in head-and-neck tumors. Int J Cancer 56:622–629

Minami K, Miki T, Kadowaki T, Seino S (2004) Roles of ATP-sensitive K$^+$ channels as metabolic sensors: studies of Kir6.x null mice. Diabetes 53 [Suppl 3]:S176–180

Mobasheri A (2002) Role of chondrocyte death and hypocellularity in ageing human articular cartilage and the pathogenesis of osteoarthritis. Med Hypotheses 58:193–197

Mobasheri A, Carter SD, Martin-Vasallo P, Shakibaei M (2002a) Integrins and stretch activated ion channels; putative components of functional cell surface mechanoreceptors in articular chondrocytes. Cell Biol Int 26:1–18

Mobasheri A, Neama G, Bell S, Richardson S, Carter SD (2002b) Human articular chondrocytes express three facilitative glucose transporter isoforms: GLUT1, GLUT3 and GLUT9. Cell Biol Int 26:297–300

Mobasheri A, Dobson H, Mason SL, Cullingham F, Shakibaei M, et al. (2005a) Expression of the GLUT1 and GLUT9 facilitative glucose transporters in embryonic chondroblasts and mature chondrocytes in ovine articular cartilage. Cell Biol Int 29:249–260

Mobasheri A, Richardson S, Mobasheri R, Shakibaei M, Hoyland JA (2005b) Hypoxia inducible factor-1 and facilitative glucose transporters GLUT1 and GLUT3: putative molecular components of the oxygen and glucose sensing apparatus in articular chondrocytes. Histol Histopathol 20:1327–1338

Mobasheri A, Vannucci SJ, Bondy CA, Carter SD, Innes JF, et al. (2002c) Glucose transport and metabolism in chondrocytes: a key to understanding chondrogenesis, skeletal development and cartilage degradation in osteoarthritis. Histol Histopathol 17: 1239–1267

Mobasheri A, Platt N, Thorpe C, Shakibaei M (2006) Regulation of 2-deoxy-D-glucose transport, lactate metabolism, and MMP-2 secretion by the hypoxia mimetic cobalt chloride in articular chondrocytes. Ann N Y Acad Sci 1091:83–93

Mobasheri A, Gent TC, Nash AI, Womack MD, Moskaluk CA, Barrett-Jolley R (2007) Evidence for functional ATP-sensitive (K(ATP)) potassium channels in human and equine articular chondrocytes. Osteoarthritis Cartilage 15:1–8

Morgello S, Uson RR, Schwartz EJ, Haber RS (1995) The human blood-brain barrier glucose transporter (GLUT1) is a glucose transporter of gray matter astrocytes. Glia 14:43–54

Moriya H, Johnston M (2004) Glucose sensing and signaling in Saccharomyces cerevisiae through the Rgt2 glucose sensor and casein kinase I. Proc Natl Acad Sci U S A 101:1572–1577

Mueckler M (1994) Facilitative glucose transporters. Eur J Biochem 219:713–725

Mueckler M, Caruso C, Baldwin SA, Panico M, Blench I, et al. (1985) Sequence and structure of a human glucose transporter. Science 229:941–945

Mueckler M, Kruse M, Strube M, Riggs AC, Chiu KC, Permutt MA (1994) A mutation in the Glut2 glucose transporter gene of a diabetic patient abolishes transport activity. J Biol Chem 269:17765–17767

Na SI, Lee MY, Heo JS, Han HJ (2007) Hydrogen peroxide increases [3H]-2-deoxyglucose uptake via MAPKs, cPLA2, and NF-kappaB signaling pathways in mouse embryonic stem cells. Cell Physiol Biochem 20:1007–1018

Nagamatsu S, Sawa H, Wakizaka A, Hoshino T (1993) Expression of facilitative glucose transporter isoforms in human brain tumors. J Neurochem 61:2048–2053

Nagano M, Yamashita T, Hamada H, Ohneda K, Kimura K, et al. (2007) Identification of functional endothelial progenitor cells suitable for the treatment of ischemic tissue using human umbilical cord blood. Blood 110:151–160

Nemoto S, Fergusson MM, Finkel T (2004) Nutrient availability regulates SIRT1 through a forkhead-dependent pathway. Science 306:2105–2108

Ohshima H, Urban JP (1992) The effect of lactate and pH on proteoglycan and protein synthesis rates in the intervertebral disc. Spine 17:1079–1082

Okma-Keulen P, Hopman-Rock M (2001) The onset of generalized osteoarthritis in older women: a qualitative approach. Arthritis Rheum 45:183–190

Otero M, Gomez Reino JJ, Gualillo O (2003) Synergistic induction of nitric oxide synthase type II: in vitro effect of leptin and interferon-gamma in human chondrocytes and ATDC5 chondrogenic cells. Arthritis Rheum 48:404–409

Otero M, Lago R, Gomez R, Dieguez C, Lago F, et al. (2006) Towards a pro-inflammatory and immunomodulatory emerging role of leptin. Rheumatology (Oxford) 45:944–950

Otero M, Lago R, Lago F, Casanueva FF, Dieguez C, et al. (2005) Leptin, from fat to inflammation: old questions and new insights. FEBS Lett 579:295–301

Otte P (1991) Basic cell metabolism of articular cartilage. Manometric studies. Z Rheumatol 50:304–312

Ouiddir A, Planes C, Fernandes I, VanHesse A, Clerici C (1999) Hypoxia upregulates activity and expression of the glucose transporter GLUT1 in alveolar epithelial cells. Am J Respir Cell Mol Biol 21:710–718

Ozcan S, Johnston M (1999) Function and regulation of yeast hexose transporters. Microbiol Mol Biol Rev 63:554–569

Pantaleon M, Kaye PL (1998) Glucose transporters in preimplantation development. Rev Reprod 3:77–81

Pantaleon M, Ryan JP, Gil M, Kaye PL (2001) An unusual subcellular localization of GLUT1 and link with metabolism in oocytes and preimplantation mouse embryos. Biol Reprod 64:1247–1254

Pao SS, Paulsen IT, Saier MH, Jr (1998) Major facilitator superfamily. Microbiol Mol Biol Rev 62:1–34

Pfander D, Cramer T, Schipani E, Johnson RS (2003) HIF-1alpha controls extracellular matrix synthesis by epiphyseal chondrocytes. J Cell Sci 116:1819–1826

Pfander D, Cramer T, Swoboda B (2005) Hypoxia and HIF-1alpha in osteoarthritis. Int Orthop 29:6–9

Phillips T, Ferraz I, Bell S, Clegg PD, Carter SD, AM (2005a) Differential regulation of the GLUT1 and GLUT3 glucose transporters by growth factors and pro-inflammatory cytokines in equine articular chondrocytes. Vet J 169:216–222

Phillips T, Ferraz I, Bell S, Clegg PD, Carter SD, Mobasheri A (2005b) Differential regulation of the GLUT1 and GLUT3 glucose transporters by growth factors and pro-inflammatory cytokines in equine articular chondrocytes. Vet J 169:216–222

Phillis JW (2004) Adenosine and adenine nucleotides as regulators of cerebral blood flow: roles of acidosis, cell swelling, and KATP channels. Crit Rev Neurobiol 16:237–270

Pickart L, Thaler MM (1980) Responses of rat and chick chondrocytes and rate hepatoma cells to plasma fractions with insulin-like and growth-promoting activities. Biochim Biophys Acta 632:112–120

Porte D, Jr., Kahn SE (1991) Mechanisms for hyperglycemia in type II diabetes mellitus: therapeutic implications for sulfonylurea treatment—an update. Am J Med 90:8S–14S

Pritzker KP (1977) Aging and degeneration in the lumbar intervertebral disc. Orthop Clin North Am 8:66–77

Pufe T, Lemke A, Kurz B, Petersen W, Tillmann B, et al. (2004) Mechanical overload induces VEGF in cartilage discs via hypoxia-inducible factor. Am J Pathol 164:185–192

Pufe T, Petersen W, Tillmann B, Mentlein R (2001) The splice variants VEGF121 and VEGF189 of the angiogenic peptide vascular endothelial growth factor are expressed in osteoarthritic cartilage. Arthritis Rheum 44:1082–1088

Quayle JM, Nelson MT, Standen NB (1997) ATP-sensitive and inwardly rectifying potassium channels in smooth muscle. Physiol Rev 77:1165–1232

Rainsford KD, Skerry TM, Chindemi P, Delaney K (1999) Effects of the NSAIDs meloxicam and indomethacin on cartilage proteoglycan synthesis and joint responses to calcium pyrophosphate crystals in dogs. Vet Res Commun 23:101–113

Rainsford KD, Ying C, Smith FC (1997) Effects of meloxicam, compared with other NSAIDs, on cartilage proteoglycan metabolism, synovial prostaglandin E2, and production of interleukins 1, 6 and 8, in human and porcine explants in organ culture. J Pharm Pharmacol 49:991–998

Rajpurohit R, Risbud MV, Ducheyne P, Vresilovic EJ, Shapiro IM (2002) Phenotypic characteristics of the nucleus pulposus: expression of hypoxia inducing factor-1, glucose transporter-1 and MMP-2. Cell Tissue Res 308:401–407

Repanti M, Korovessis PG, Stamatakis MV, Spastris P, Kosti P (1998) Evolution of disc degeneration in lumbar spine: a comparative histological study between herniated and postmortem retrieved disc specimens. J Spinal Disord 11:41–45

Richardson DC, Zentek J (1998) Nutrition and osteochondrosis. Vet Clin North Am Small Anim Pract 28:115–135

Richardson DW, Dodge GR (2000) Effects of interleukin-1beta and tumor necrosis factor-alpha on expression of matrix-related genes by cultured equine articular chondrocytes. Am J Vet Res 61:624–630

Richardson S, Neama G, Phillips T, Bell S, Carter SD, et al. (2003) Molecular characterization and partial cDNA cloning of facilitative glucose transporters expressed in human articular chondrocytes; stimulation of 2-deoxyglucose uptake by IGF-I and elevated MMP-2 secretion by glucose deprivation. Osteoarthritis Cartilage 11:92–101

Risbud MV, Albert TJ, Guttapalli A, Vresilovic EJ, Hillibrand AS, et al. (2004a) Differentiation of mesenchymal stem cells towards a nucleus pulposus-like phenotype in vitro: implications for cell-based transplantation therapy. Spine 29:2627–2632

Risbud MV, Shapiro IM, Vaccaro AR, Albert TJ (2004b) Stem cell regeneration of the nucleus pulposus. Spine J 4:348S–53S

Risbud MV, Izzo MW, Adams CS, Arnold WW, Hillibrand AS, et al. (2003) An organ culture system for the study of the nucleus pulposus: description of the system and evaluation of the cells. Spine 28:2652–2658; discussion 8–9

Robbins JR, Thomas B, Tan L, Choy B, Arbiser JL, et al. (2000) Immortalized human adult articular chondrocytes maintain cartilage-specific phenotype and responses to interleukin-1beta. Arthritis Rheum 43:2189–2201

Roberts S (2002) Disc morphology in health and disease. Biochem Soc Trans 30:864–869

Rolland F, Moore B, Sheen J (2002) Sugar sensing and signaling in plants. Plant Cell 14 [Suppl]:S185–205

Rolland F, Winderickx J, Thevelein JM (2001) Glucose-sensing mechanisms in eukaryotic cells. Trends Biochem Sci 26:310–317

Rosenbloom AL, Silverstein JH (1996) Connective tissue and joint disease in diabetes mellitus. Endocrinol Metab Clin North Am 25:473–483

Roughley PJ (2006) The structure and function of cartilage proteoglycans. Eur Cell Mater 12:92–101

Saier MH, Jr., Beatty JT, Goffeau A, Harley KT, Heijne WH, et al. (1999a) The major facilitator superfamily. J Mol Microbiol Biotechnol 1:257–279

Saier MH, Jr., Eng BH, Fard S, Garg J, Haggerty DA, et al. (1999b) Phylogenetic characterization of novel transport protein families revealed by genome analyses. Biochim Biophys Acta 1422:1–56

Saier MH, Jr., Paulsen IT, Sliwinski MK, Pao SS, Skurray RA, Nikaido H (1998) Evolutionary origins of multidrug and drug-specific efflux pumps in bacteria. FASEB J 12:265–274

Santalucia T, Camps M, Castello A, Munoz P, Nuel A, et al. (1992) Developmental regulation of GLUT-1 (erythroid/Hep G2) and GLUT-4 (muscle/fat) glucose transporter expression in rat heart, skeletal muscle, and brown adipose tissue. Endocrinology 130:837–846

Santer R, Groth S, Kinner M, Dombrowski A, Berry GT, et al. (2002) The mutation spectrum of the facilitative glucose transporter gene SLC2A2 (GLUT2) in patients with Fanconi-Bickel syndrome. Hum Genet 110:21–29

Scheepers A, Joost HG, Schurmann A (2004) The glucose transporter families SGLT and GLUT: molecular basis of normal and aberrant function. JPEN J Parenter Enteral Nutr 28:364–371

Schipani E, Ryan HE, Didrickson S, Kobayashi T, Knight M, Johnson RS (2001) Hypoxia in cartilage: HIF-1alpha is essential for chondrocyte growth arrest and survival. Genes Dev 15:2865–2876

Schwartz ER, Adamy L (1977) Effect of ascorbic acid on arylsulfatase activities and sulfated proteoglycan metabolism in chondrocyte cultures. J Clin Invest 60:96–106

Schwartz ER, Oh WH, Leveille CR (1981) Experimentally induced osteoarthritis in guinea pigs: metabolic responses in articular cartilage to developing pathology. Arthritis Rheum 24:1345–1355

Seatter MJ, Gould GW (1999) The mammalian facilitative glucose transporter (GLUT) family. Pharm Biotechnol 12:201–228

Semenza G (2002a) Signal transduction to hypoxia-inducible factor 1. Biochem Pharmacol 64:993–998

Semenza GL (1998) Hypoxia-inducible factor 1: master regulator of O2 homeostasis. Curr Opin Genet Dev 8:588–594

Semenza GL (1999a) Perspectives on oxygen sensing. Cell 98:281–284

Semenza GL (1999b) Regulation of mammalian O2 homeostasis by hypoxia-inducible factor 1. Annu Rev Cell Dev Biol 15:551–578

Semenza GL (2001) Regulation of hypoxia-induced angiogenesis: a chaperone escorts VEGF to the dance. J Clin Invest 108:39–40

Semenza GL (2002b) HIF-1 and tumor progression: pathophysiology and therapeutics. Trends Mol Med 8:S62–67

Semenza GL, Agani F, Booth G, Forsythe J, Iyer N, et al. (1997) Structural and functional analysis of hypoxia-inducible factor 1. Kidney Int 51:553–555

Semenza GL, Agani F, Iyer N, Kotch L, Laughner E, et al. (1999) Regulation of cardiovascular development and physiology by hypoxia-inducible factor 1. Ann NY Acad Sci 874:262–268

Semenza GL, Artemov D, Bedi A, Bhujwalla Z, Chiles K, et al. (2001) 'The metabolism of tumours': 70 years later. Novartis Found Symp 240:251–260; discussion 260–264

Semenza GL, Jiang BH, Leung SW, Passantino R, Concordet JP, et al. (1996) Hypoxia response elements in the aldolase A, enolase 1, and lactate dehydrogenase A gene promoters contain essential binding sites for hypoxia-inducible factor 1. J Biol Chem 271:32529–32537

Semenza GL, Roth PH, Fang HM, Wang GL (1994) Transcriptional regulation of genes encoding glycolytic enzymes by hypoxia-inducible factor 1. J Biol Chem 269:23757–23763

Semenza GL, Wang GL (1992) A nuclear factor induced by hypoxia via de novo protein synthesis binds to the human erythropoietin gene enhancer at a site required for transcriptional activation. Mol Cell Biol 12:5447–5454

Shakibaei M, John T, Schulze-Tanzil G, Lehmann I, Mobasheri A (2007a) Suppression of NF-kappaB activation by curcumin leads to inhibition of expression of cyclo-oxygenase-2 and matrix metalloproteinase-9 in human articular chondrocytes: Implications for the treatment of osteoarthritis. Biochem Pharmacol 73:1434–1445

Shakibaei M, John T, Seifarth C, Mobasheri A (2007b) Resveratrol inhibits IL-1beta-induced stimulation of caspase-3 and cleavage of PARP in human articular chondrocytes in vitro. Ann NY Acad Sci 1095:554–563

Shakibaei M, Kociok K, Forster C, Vormann J, Gunther T, et al. (1996) Comparative evaluation of ultrastructural changes in articular cartilage of ofloxacin-treated and magnesium-deficient immature rats. Toxicol Pathol 24:580–587

Shakibaei M, Schulze-Tanzil G, John T, Mobasheri A (2005) Curcumin protects human chondrocytes from IL-l1beta-induced inhibition of collagen type II and beta1-integrin expression and activation of caspase-3: an immunomorphological study. Ann Anat 187:487–497

Shepherd PR, Gould GW, Colville CA, McCoid SC, Gibbs EM, Kahn BB (1992) Distribution of GLUT3 glucose transporter protein in human tissues. Biochem Biophys Res Commun 188:149–154

Shikhman AR, Brinson DC, Lotz MK (2004) Distinct pathways regulate facilitated glucose transport in human articular chondrocytes during anabolic and catabolic responses. Am J Physiol Endocrinol Metab 286:E980–985

Shikhman AR, Brinson DC, Valbracht J, Lotz MK (2001a) Cytokine regulation of facilitated glucose transport in human articular chondrocytes. J Immunol 167:7001–7008

Shikhman AR, Kuhn K, Alaaeddine N, Lotz M (2001b) N-acetylglucosamine prevents IL-1 beta-mediated activation of human chondrocytes. J Immunol 166:5155–5160

Shin BC, McKnight RA, Devaskar SU (2004) Glucose transporter GLUT8 translocation in neurons is not insulin responsive. J Neurosci Res 75:835–844

Skawina A, Litwin JA, Gorczyca J, Miodonski AJ (1994) Blood vessels in epiphyseal cartilage of human fetal femoral bone: a scanning electron microscopic study of corrosion casts. Anat Embryol (Berl) 189:457–462

Slater MR, Scarlett JM, Donoghue S, Kaderly RE, Bonnett BN, et al. (1992) Diet and exercise as potential risk factors for osteochondritis dissecans in dogs. Am J Vet Res 53:2119–2124

Smith KJ, Bertone AL, Weisbrode SE, Radmacher M (2006) Gross, histologic, and gene expression characteristics of osteoarthritic articular cartilage of the metacarpal condyle of horses. Am J Vet Res 67:1299–1306

Stahlmann R, Forster C, Shakibaei M, Vormann J, Gunther T, Merker HJ (1995) Magnesium deficiency induces joint cartilage lesions in juvenile rats which are identical to quinolone-induced arthropathy. Antimicrob Agents Chemother 39:2013–2018

Stahlmann R, Kuhner S, Shakibaei M, Flores J, Vormann J, van Sickle DC (2000) Effects of magnesium deficiency on joint cartilage in immature beagle dogs: immunohistochemistry, electron microscopy, and mineral concentrations. Arch Toxicol 73:573–580

Stavrou S, Kleinberg DL (2001) Rheumatic manifestations of pituitary tumors. Curr Rheumatol Rep 3:459–463

Studer RK, Georgescu HI, Miller LA, Evans CH (1999) Inhibition of transforming growth factor beta production by nitric oxide-treated chondrocytes: implications for matrix synthesis. Arthritis Rheum 42:248–257

Sztrolovics R, White RJ, Poole AR, Mort JS, Roughley PJ (1999) Resistance of small leucine-rich repeat proteoglycans to proteolytic degradation during interleukin-1-stimulated cartilage catabolism. Biochem J 339 (Pt 3):571–577

Takakura Y, Kuentzel SL, Raub TJ, Davies A, Baldwin SA, Borchardt RT (1991) Hexose uptake in primary cultures of bovine brain microvessel endothelial cells. I. Basic characteristics and effects of D-glucose and insulin. Biochim Biophys Acta 1070:1–10

Tal M, Schneider DL, Thorens B, Lodish HF (1990) Restricted expression of the erythroid/brain glucose transporter isoform to perivenous hepatocytes in rats. Modulation by glucose. J Clin Invest 86:986–992

Thompson RE, Pearcy MJ, Downing KJ, Manthey BA, Parkinson IH, Fazzalari NL (2000) Disc lesions and the mechanics of the intervertebral joint complex. Spine 25:3026–3035

Thorens B (1996) Glucose transporters in the regulation of intestinal, renal, and liver glucose fluxes. Am J Physiol 270:G541–553

Thorens B, Cheng ZQ, Brown D, Lodish HF (1990) Liver glucose transporter: a basolateral protein in hepatocytes and intestine and kidney cells. Am J Physiol 259:C279–285

Todhunter PG, Kincaid SA, Todhunter RJ, Kammermann JR, Johnstone B, et al. (1996) Immunohistochemical analysis of an equine model of synovitis-induced arthritis. American Journal of Veterinary Research 57:1080–1093

Tonack S, Rolletschek A, Wobus AM, Fischer B, Santos AN (2006) Differential expression of glucose transporter isoforms during embryonic stem cell differentiation. Differentiation 74:499–509

Torzilli PA, Arduino JM, Gregory JD, Bansal M (1997) Effect of proteoglycan removal on solute mobility in articular cartilage. J Biomech 30:895–902

Torzilli PA, Grande DA, Arduino JM (1998) Diffusive properties of immature articular cartilage. J Biomed Mater Res 40:132–138

Trippel SB (1995) Growth factor actions on articular cartilage. J Rheumatol Suppl 43:129–132

Uldry M, Thorens B (2004) The SLC2 family of facilitated hexose and polyol transporters. Pflugers Arch 447:480–489

Underwood LE (1996) Nutritional regulation of IGF-I and IGFBPs. J Pediatr Endocrinol Metab 9 [Suppl 3]:303–312

Urban JP, Roberts S (2003) Degeneration of the intervertebral disc. Arthritis Res Ther 5:120–130

Urban MR, Fairbank JC, Etherington PJ, Loh FL, Winlove CP, Urban JP (2001) Electrochemical measurement of transport into scoliotic intervertebral discs in vivo using nitrous oxide as a tracer. Spine 26:984–990

Vachon AM, Keeley FW, McIlwraith CW, Chapman P (1990) Biochemical analysis of normal articular cartilage in horses. Am J Vet Res 51:1905–1911

van den Berg WB (1999) The role of cytokines and growth factors in cartilage destruction in osteoarthritis and rheumatoid arthritis. Z Rheumatol 58:136–141

Vannucci RC, Vannucci SJ (2000) Glucose metabolism in the developing brain. Semin Perinatol 24:107–115

Vannucci SJ (1994) Developmental expression of GLUT1 and GLUT3 glucose transporters in rat brain. J Neurochem 62:240–246

Vannucci SJ, Reinhart R, Maher F, Bondy CA, Lee WH, et al. (1998) Alterations in GLUT1 and GLUT3 glucose transporter gene expression following unilateral hypoxia-ischemia in the immature rat brain. Brain Res Dev Brain Res 107:255–264

Vannucci SJ, Seaman LB, Vannucci RC (1996) Effects of hypoxia-ischemia on GLUT1 and GLUT3 glucose transporters in immature rat brain. J Cereb Blood Flow Metab 16:77–81

Virgintino D, Robertson D, Monaghan P, Errede M, Bertossi M, et al. (1997) Glucose transporter GLUT1 in human brain microvessels revealed by ultrastructural immunocytochemistry. J Submicrosc Cytol Pathol 29:365–370

Vogel KG, Paulsson M, Heinegard D (1984) Specific inhibition of type I and type II collagen fibrillogenesis by the small proteoglycan of tendon. Biochem J 223:587–597

von der Mark K, Kirsch T, Nerlich A, Kuss A, Weseloh G, et al. (1992) Type X collagen synthesis in human osteoarthritic cartilage. Indication of chondrocyte hypertrophy. Arthritis Rheum 35:806–811

Waeber G, Pedrazzini T, Bonny O, Bonny C, Steinmann M, et al. (1995) A 338-bp proximal fragment of the glucose transporter type 2 (GLUT2) promoter drives reporter gene expression in the pancreatic islets of transgenic mice. Mol Cell Endocrinol 114:205–215

Wang GL, Jiang BH, Rue EA, Semenza GL (1995a) Hypoxia-inducible factor 1 is a basic-helix-loop-helix-PAS heterodimer regulated by cellular O2 tension. Proc Natl Acad Sci USA 92:5510–5514

Wang GL, Jiang BH, Semenza GL (1995b) Effect of altered redox states on expression and DNA-binding activity of hypoxia-inducible factor 1. Biochem Biophys Res Commun 212:550–556

Wang J, Zhou J, Bondy CA (1999) Igf1 promotes longitudinal bone growth by insulin-like actions augmenting chondrocyte hypertrophy. FASEB J 13:1985–1990

Watanabe H, Yamada Y, Kimata K (1998) Roles of aggrecan, a large chondroitin sulfate proteoglycan, in cartilage structure and function. J Biochem (Tokyo) 124:687–693

Watson RT, Pessin JE (2001) Intracellular organization of insulin signaling and GLUT4 translocation. Recent Prog Horm Res 56:175–193

Westacott CI, Sharif M (1996) Cytokines in osteoarthritis: mediators or markers of joint destruction? Semin Arthritis Rheum 25:254–272

Wiberg C, Klatt AR, Wagener R, Paulsson M, Bateman JF, et al. (2003) Complexes of matrilin-1 and biglycan or decorin connect collagen VI microfibrils to both collagen II and aggrecan. J Biol Chem 278:37698–37704

Wilhelmi G (1993a) [Potential effects of nutrition including additives on healthy and arthrotic joints. I. Basic dietary constituents]. Z Rheumatol 52:174–179

Wilhelmi G (1993b) [Potential influence of nutrition with supplements on healthy and arthritic joints. II. Nutritional quantity, supplements, contamination]. Z Rheumatol 52:191–200

Williams JS, Jr., Bush-Joseph CA, Bach BR, Jr (1998) Osteochondritis dissecans of the knee. Am J Knee Surg 11:221–232

Wolf HJ, Desoye G (1993) Immunohistochemical localization of glucose transporters and insulin receptors in human fetal membranes at term. Histochemistry 100:379–385

Wood IS, Trayhurn P (2003) Glucose transporters (GLUT and SGLT): expanded families of sugar transport proteins. Br J Nutr 89:3–9

Woodard JC, Becker HN, Poulos PW, Jr (1987a) Articular cartilage blood vessels in swine osteochondrosis. Vet Pathol 24:118–123

Woodard JC, Becker HN, Poulos PW, Jr (1987b) Effect of diet on longitudinal bone growth and osteochondrosis in swine. Vet Pathol 24:109–117

Woodbury MR, Feist MS, Clark EG, Haigh JC (1999) Osteochondrosis and epiphyseal bone abnormalities associated with copper deficiency in bison calves. Can Vet J 40:878–880

Wu X, Freeze HH (2002) GLUT14, a duplicon of GLUT3, is specifically expressed in testis as alternative splice forms. Genomics 80:553–557

Wyman AH, Chi M, Riley J, Carayannopoulos MO, Yang C, et al. (2003) Syntaxin 4 expression affects glucose transporter 8 translocation and embryo survival. Mol Endocrinol 17:2096–2102

Yamaguchi T, Hayashi K, Tayama N, Sugioka Y (1990) The role of cartilage canals: experimental study using rabbits' femoral heads. Nippon Seikeigeka Gakkai Zasshi 64:1105–1110

Yasuma T, Koh S, Okamura T, Yamauchi Y (1990) Histological changes in aging lumbar intervertebral discs. Their role in protrusions and prolapses. J Bone Joint Surg Am 72:220–229

Ybarra J, Behrooz A, Gabriel A, Koseoglu MH, Ismail-Beigi F (1997) Glycemia-lowering effect of cobalt chloride in the diabetic rat: increased GLUT1 mRNA expression. Mol Cell Endocrinol 133:151–160

Yosimichi G, Nakanishi T, Nishida T, Hattori T, Takano-Yamamoto T, Takigawa M (2001) CTGF/Hcs24 induces chondrocyte differentiation through a p38 mitogen-activated protein kinase (p38MAPK), and proliferation through a p44/42 MAPK/extracellular-signal regulated kinase (ERK). Eur J Biochem 268:6058–6065

Younes M, Lechago LV, Somoano JR, Mosharaf M, Lechago J (1996) Wide expression of the human erythrocyte glucose transporter Glut1 in human cancers. Cancer Res 56:1164–1167

Yu S, Tooyama I, Ding WG, Kitasato H, Kimura H (1995) Immunohistochemical localization of glucose transporters (GLUT1 and GLUT3) in the rat hypothalamus. Obes Res 3 [Suppl 5]:753S–76S

Zhang JZ, Behrooz A, Ismail-Beigi F (1999) Regulation of glucose transport by hypoxia. Am J Kidney Dis 34:189–202

Index

Made in the USA
Monee, IL
07 July 2026

56546130R00057